Planetary Exploration with Ingenuity and Dragonfly

Rotary-Wing Flight on Mars and Titan

Planetary Exploration with Ingenuity and Dragonfly

Rotary-Wing Flight on Mars and Titan

Ralph D. Lorenz

Paul Park, Editor-in-Chief
University of Texas at Arlington
Arlington, Texas

Published by
American Institute of Aeronautics and Astronautics, Inc.
12700 Sunrise Valley Drive, Reston, VA 20191-5807

Cover image: Ingenuity Helicopter on Mars (Illustration)
Credit: NASA/JPL-Caltech

American Institute of Aeronautics and Astronautics, Inc., Reston, Virginia

1 2 3 4 5

Library of Congress Cataloging-in-Publication Data

ISBN: 978-1-62410-636-1

Ebook ISBN: 978-1-62410-637-8

CONTENTS

PREFACE

The idea for this book emerged when I was in the company of astronauts, space artists, and others at Spacefest in Tucson in summer 2021. I had given a talk about the Dragonfly mission, and as was often the case, a number of the audience questions afterward had asked about lessons from the Ingenuity helicopter then flying on Mars. I realized two things. First, that a book outlining the very different scope and origin of the two projects, the very different vehicle designs, and the very different environments they fly in would be useful. And second, that I would be uniquely well-placed to write such a book, as will (I hope) become clear.

There is something of an irony in my writing a book about aeronautics. I was fortunate in my teens to realize what it was I wanted to do professionally—to work on robotic planetary exploration. An undergraduate degree in Aerospace Systems Engineering was an obvious direction for me to take, and in the later years of the course included the classes in astronautical topics that I really wanted to pursue, such as orbital mechanics. In the meantime, though, I had to sit through the obligatory traditional aeronautics topics such as aerodynamics. There was even a class on helicopter engineering, which I found surprisingly interesting, although I never expected to use its material professionally.

My first job after my degree saw me realizing my dream—I worked at the European Space Agency on the design and early development of the Huygens probe to Titan. Although it did indeed fly in space, it also flew* in an atmosphere, and so the subjects I endured turned out to be useful after all!

My work at ESA centered on the accommodation of the scientific instruments on the probe and on various aspects of the environment that the probe would have to endure. This set me on a path toward planetary science and all its subdisciplines such as geology and meteorology, to the point where if I am now asked my profession, I tend to say *planetary scientist*; however, in reality, my day-to-day contribution is more one of impedance matching between scientists and engineers.

---

*if "descending by parachute" counts as flying....

Such a translator role is a big part of my activity on the Dragonfly project, where many colleagues have experience working on spacecraft but have little familiarity with planetary atmospheres or surfaces, or with aeronautical systems. And hence this book, which aims to lay out to "space people" what they need to know about rotorcraft and to "helicopter people" what they need to know about delivering vehicles through space and operating them on other planets.

This book cannot be a formal textbook on rotorcraft, spacecraft, or engineering. But it will, I hope, outline the basics of these distinct endeavors and their intersections. I have provided references for the interested reader to delve further. The book is not quite a history either, but more of a snapshot: Ingenuity's story is not quite finished, and much of Dragonfly's story lies in the future. This has confronted me with the question of "if not now, then when?" It seemed to me that there would be value in a synthesis of the state of the art today, even if incomplete.

The providers of graphic material are noted in the relevant captions, but let me record here my appreciation for their contribution. DC Agle at Jet Propulsion Laboratory facilitated the sourcing of material on Ingenuity testing. I am most grateful to Matt Keennon, Ben Pipenberg, Sara Langberg, and Jeremy Tyler for a congenial and productive visit to AeroVironment, Inc. to make measurements on the terrestrial demonstrator ("Terry") of the Mars helicopter. I appreciate insightful discussions with Bob Balaram, Håvard Grip and Teddy Tzanetos of the JPL Ingenuity team. I also recognize the valuable discussions I had with my Supercam colleagues in France on the interpretation of the sound of the Ingenuity helicopter recorded by that instrument on Mars. Larry Young, of NASA Ames, and Anubhav Datta, of the University of Maryland, shared useful perspectives on the prehistory of the Mars helicopter.

It is a tremendous privilege to work with a talented (and growing) team on Dragonfly. Several Dragonfly colleagues were kind enough to read parts or all of the manuscript and offer comments and fact-check: Karen Kirby, Rick Fitzgerald, and especially Jack Langelaan. Teddy Tzanetos was kind enough to fact-check the chapters on Ingenuity, and to provide an eloquent Foreword. Any errors, however, are the responsibility of the author.

Ralph Lorenz
Columbia, Maryland, November 2021

FOREWORD

The Montgolfière "Le Réveillon" (1783, first hot-air balloon), the Wright Flyer (1903, first airplane), the Vought-Sikorsky VS-300 (1939, first main-rotor helicopter), the Ingenuity Mars Helicopter (2021, first rotorcraft at Mars), and Dragonfly (sched. circa 2034, first rotorcraft at Saturn's moon Titan) each embody humanity's pioneering spirit to reach for the skies. The Ingenuity and Dragonfly teams are expanding Orville's and Wilbur's vision to the very limits of engineering and physics to make extra-terrestrial flight a reality. Each celestial body presents its own unique set of challenges to over-come. For Ingenuity, flight in Mars' thin atmosphere (~2% density of Earth's) is only possible with an extremely lightweight and efficient vehicle aircraft weighing only 1.8kg (3.96lbs). Titan's bitter cold (-179 deg. C/-290 deg. F) and distance from Earth make designing an autonomous rotorcraft like Dragonfly a revolutionary step for flight in other ways.

Most of Ingenuity's pre-launch milestones revolved around proving first-of-kinds demonstrations of aircraft capability in a Mars-like environment. For example, proving basic lift generation in a Mars-representative volume of air, demonstrating autonomous navigation using on-board avionics, and surviving its first simulated Martian night-sol cycle in a Mars-like thermal chamber were critical milestones in establishing Ingenuity's flight-worthiness. Each test was a small step towards the launchpad, asserting that the dream of flight on Mars was that much closer to reality.

Ingenuity was built as a NASA Technology Demonstration aboard the Mars 2020 Perseverance mission. Although it has greatly outperformed its initial 30-sol mission on Mars and became a forward-scout for the Perseverance rover, Ingenuity carries no scientific payloads onboard. In contrast, Dragonfly is part of NASA's New Frontiers Program, and will bring the miracle of flight to a Saturn's moon Titan where it will deploy of bevy of scientific payloads to study Titan's habitability and prebiotic chemistry. Titan is an ocean world and the only moon in our solar system with a dense atmosphere, which supports an Earth-like hydrological cycle of methane clouds, rain, and liquid flowing across the surface to fill lakes and seas.

An important distinction from aircraft designed for Earth, is that platforms like Ingenuity and Dragonfly are not only aircraft, but also fully-fledged spacecraft. Like most spacecraft, they need to support a specific concept of operations, command and control capabilities, telecommunications with operators back at Earth, thermal management, fault management, scientific and engineering data collection and storage, and most critically survival in their harsh environments. This text introduces the concept of planetary rotorcraft and the myriad technical aspects of creating and operating extra-terrestrial aircraft, and serves as a primer on the Ingenuity and Dragonfly missions. For readers familiar with spacecraft design, this book will serve as an introduction to rotorcraft design and flight, and vice-versa.

Just as the Wright Flyer ushered in a pioneering century of aviation on Earth, the Ingenuity and Dragonfly teams hope to similarly be catalysts for future fleets of extra-terrestrial aircraft on Mars, Titan, and beyond. This book endeavors to explain the exciting challenges of extra-terrestrial flight, and provide a glimpse into the budding future of flight on other worlds.

Theodore "Teddy" Tzanetos
Ingenuity Team Lead
NASA Jet Propulsion Laboratory[†]
California Institute of Technology

[†]This work was carried out at the Jet Propulsion Laboratory, California Institute of Technology, under a contract with the National Aeronautics and Space Administration (80NM0018D0004).

Chapter 1

ORIGINS OF PLANETARY FLIGHT

"Given ships or sails adapted to the breezes of heaven, there will be those who will not shrink from even that vast expanse."

—Johannes Kepler, 1610

Flight is one of the most profound aspirations of humans. The means by which to attain it was not obvious, however, and in the late 19th and early 20th centuries, developments were pursued for balloons, airships, airplanes, helicopters, and indeed rockets. The emergence of practical vehicles for different applications only took place beginning in World War I. In this chapter, we recap a brief history of rotorcraft development,[1] in parallel with the progress of planetary science, to explain how Ingenuity and Dragonfly emerged.

A toy hand-spun rotor, usually made from bamboo, existed in China from around 300 AD and appears in Renaissance paintings in Europe. The extension of the principle to person-carrying flight was famously imagined by Leonardo da Vinci, who sketched his famous, but impractical, "aerial screw" in the late 1480s (see Fig. 1.1).

Perhaps the first modern calculation[2] on how lift might be generated to support a human in flight was performed by a scientist more famous for his contributions in astronomy—Edmond Halley. In a not widely known report from 1691, he had a servant procure a pigeon, which he then suspended in a channel of flowing water. Halley was familiar with Newton's foundational work on mechanics (*Principia*)—indeed he had arranged for the publication of it—so he knew that the impetus, or force, on a wing should vary as the product of its area, the density of the fluid, and the square of the fluid speed. This concept, the idea of force as a flux of momentum, was the fundamental basis for estimating lift. Halley could then deduce:

1

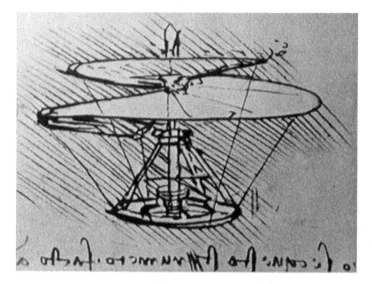

Fig. 1.1   Da Vinci's aerial screw, imagined to be made of linen.

> Wings large enough to sustain the weight of one man with the same veloc-
> ity of the stroke ought to be in Area 200 times as large as those of this
> bird, that is in their linear dimensions 14 times as great and consequently
> they ought to be about 12 foot long.

This result is not too different from a modern hang glider. This type of momentum calculation is central to assessing the practicability of flight in other atmospheres.

Flight on other worlds was contemplated by the Dutch polymath Christiaan Huygens in his posthumous *The Celestial Worlds Discover'd* (1698). Huygens, who discovered Saturn's moon Titan in 1655, imagined that other planetary bodies might have atmospheres of a constitution and density quite different from our own, with possibly favorable implications for aerial locomotion.* Huygens, however, knew the scale of the solar system and calculated that a bullet—the fastest thing he knew about—would take 25 years to travel from the sun to the Earth.

Interestingly, the rotor flight–planetary science connection dates surprisingly far back.[3] In July 1754, Russian Mikhail Lomonosov demonstrated a small clockwork, spring-driven coaxial helicopter to the Russian Academy of Sciences (see Fig. 1.2). The contrarotating fans solved the problem of the drag torque on a single rotor causing the platform to spin unhelpfully. He proposed the device as a means to loft meteorological instrumentation to high altitudes. Lomonosov was also the

---

*"The Air may be I confess much thicker and heavier than ours, and so without disadvantage to its transparency, be fitter for the volatile animals."

discoverer of Venus's atmosphere, observing how it refracted light during the transit of Venus across the face of the sun in 1761.

Practical aeronautics had its beginnings at this time, with the 1783 piloted flight by Pilatre de Rozier in a hot-air balloon built by the Montgolfier brothers, and the flight just 10 days later of a completely different type of balloon (hydrogen-filled) by Jacques Charles.

In 1792, Englishman George Cayley began experimenting with helicopter tops, which he later called *rotary wafts* or *elevating fliers*. His landmark 1809 article, "On Aerial Navigation," considers the feasibility of flight. To help familiarize his readers with the practicalities, he described a flying model with two propellers (constructed from corks and feathers) powered by a whalebone leaf spring arranged as a bow drill to spin the rotors. This machine reached heights of a few tens of feet. He proposed that a scaled-up version, with 200 ft^2 rotors, could carry a person, but noted, "for the mere purpose of ascent this is perhaps the best apparatus; but speed ... requires a different structure."

This last remark highlights a fundamental distinction between fixed-wing and moving-wing flight—the former may be better for speed, whereas moving wings are more practical for taking off. Although the power-to-weight ratio of elastic energy storage such as whalebone springs is adequate for flight, the

Fig. 1.2   Reconstruction of Lomonosov's 1754 clockwork, spring-driven instrument platform, which used contrarotating coaxial rotors.
Source: CC-BY-SA, Albina-Belenkaya, Sept. 2014.

energy-to-weight ratio is very poor, so demonstrations like this were limited to less than a minute. Cayley's note recognizes the supreme importance of these considerations, and that although the coal consumption of the steam engines of the day was suitably modest for flight, the engines themselves were too heavy. [Thus, whereas the specific energy of steam engines and their fuel (J/kg) was high, their specific power (W/kg) was not.] Presciently, Cayley anticipated the petroleum-fueled internal combustion engine as being the solution:

> It is proper to notice the probability that exists of using the expansion of air by the sudden combustion of inflammable powders or fluids with great advantage. The French have lately shown the great power produced by igniting inflammable powders in close vessels; and several years ago an engine was made to work in this country in a similar manner, by the inflammation of spirit of tar.

The subsequent development of practical aeronautics focused on fixed-wing flight, and the efforts in this direction of Cayley, Otto Lilienthal, Samuel Pierpoint Langley, Octave Chanute, and Wilbur and Orville Wright have been well documented. Among these aeronautic pioneers, Langley also had interests that pertained to planetary atmospheres: He helped develop thermopile detectors for telescopes, by which the radiant heat (infrared light) of the moon or planets could be measured to remotely estimate their temperatures.

Meanwhile, the French inventor Gustave Ponton d'Amécourt built a steam-powered rotorcraft in 1861 (see Fig. 1.3), attempting to get around the power-to-weight problem by using the then-new wonderfully lightweight metal aluminum. (In fact, his contraption may be the earliest surviving mechanical artifact to have been built from this material.) Although his invention did not get into the air, the name he gave it, *hélicoptère*, did prove to be a lasting contribution to the field (perhaps aided by its use in a Jules Verne story). The first part of the word, *helico*, derives from the Greek word *helix*, meaning spiral or whirl; the second part, *pter* derives from *pteron*, meaning wing. It is interesting that a century and a half later, the word *helicopter* has been split in a different place from the original join to give *quadcopter*, *octocopter*, and the like.

Various inventors in the subsequent decades, including Thomas Edison, experimented with model rotorcraft powered by steam and gasoline engines, and by electric motors. Extension to human-carrying scale was achieved in the early 1900s. As is usual in history, exactly what can be called the "first helicopter flight" depends on the benchmark chosen—a flight in Berlin in 1901 may be the first, but is not well documented. Meanwhile, Jacques and Louis Breguet in France developed a quad-rotor gyroplane that flew unsteadily in summer 1907 for 20 s, but it was tethered to the ground.

What is often considered the first helicopter flight was made by Frenchman Paul Cornu, who like the Wright brothers was a bicycle manufacturer. His vehicle had two 6-m-diameter contrarotating rotors on either side of a frame

Fig. 1.3 The original helicoptère, on display at the museum in Le Bourget outside Paris. The coaxial device—note that the lower rotor blades are seen end-on here—would have flown if the steam engine were only lighter.
Source: Author.

that contained the pilot and an 18-kW (24 hp) gasoline engine. Flights up to a meter or two, for an unsteady minute or so, were accomplished, although the poor controllability led to the abandonment of the design.

By the 1920s, helicopters were beginning to look like practical vehicles. It is interesting to recall that the first practical helicopter to fly in the United States, in 1924, was a quad-rotor vehicle, the de Bothezat "flying octopus" (see Fig. 1.4). Although this vehicle flew more than 100 times with as many as four passengers and broke many records, the pilot workload to achieve control by differential thrust on four rotors, each with variable pitch, was formidable. Although the same capabilities were not reproduced for another 20 years, the Army Air Service scrapped the project.

Around this time, an Argentinian working in Europe, Raúl Castelluccio, introduced the innovation of cyclic pitch. By changing the angle of a rotor blade over the course of each revolution, side forces could be generated without requiring the whole rotor disk to be moved and without requiring separate maneuvering fans. This allowed the control problem to be solved without requiring three or more rotors. Castelluccio's helicopter Number 3 flew for up to 10 min.

Progress in planetary astronomy was being made at this time as well. In the 1920s, observations of the absorption and reflection of light at different

Fig. 1.4   The de Bothezat quadcopter or "flying octopus," the most successful rotorcraft in the United States for almost 20 years. Notice the cylindrical frames around the rotor hubs and the small fan for yaw control at left.
Source: U.S. Army Air Corps, digitized by author at the National Archives, Silver Spring, Md.

wavelengths (spectroscopy) on Mars, and consideration of its temperature indicated in the infrared, opened the way to attempts to estimate how thick its atmosphere might be. The planets were beginning to be quantified places rather than merely little circles in the sky.

After something of a digression in the 1920s and 1930s with the autogiro, an odd hybrid vehicle with an unpowered rotor, practical helicopters emerged.[†] In Germany in 1938, Heinrich Focke's Fw-61 set an altitude record of some 11,000 ft (3.5 km) and was famously filmed making precision indoor flights

---

[†]Arguably another digression of the period was the airship. The Graf Zeppelin and Hindenburg were operating regular transatlantic passenger flights until the infamous Hindenburg disaster in 1937.

in Berlin at the hands of Hanna Reitsch. The Fw-61 had two side-by-side rotors that were derived from those originally developed for autogiros. A larger helicopter of similar configuration (the Fa-223 Drache, "Dragon") was used experimentally during World War II to carry cargo loads slung underneath, a mission commonly performed by helicopters since. A more compact configuration, with two slightly canted intermeshing rotors, was used in a small German shipborne reconnaissance helicopter, the Flettner Fl 282 Kolibri ("Hummingbird"). This is considered the first series production helicopter, although Allied bombing of the factory limited the number put into service from 1000 to 24.

Meanwhile in the United States, Fw-61–like designs were being pursued in parallel with what would become the standard configuration of a single large rotor, with the reaction torque compensated by a small rotor on a tail boom. Variants of this arrangement were developed by Arthur Young for the Bell Aircraft company and by Igor Sikorsky for his company (see Fig. 1.5). By the end of the war more than 100 Sikorsky R-4s would be in service for transport and rescue, and the 1945 Bell 47 design would become the first helicopter certified for civilian use and would sell some 5600 airframes. The helicopter had arrived.

Fig. 1.5 Like many early rotorcraft innovators, Igor Sikorsky had to be an entrepreneur and showman to get his inventions off the ground commercially as well as technically. Here in 1944 he demonstrates the use of a winch on the R-4 for rescue applications. Source: U.S. Coast Guard photo.

THE SPACE AGE

World War II brought myriad technical developments, in computers, radio systems, guidance, nuclear technology, and of course of rocket propulsion, that would make interplanetary travel possible within two and a half decades. In 1944, Dutch astronomer Gerard Kuiper made an observation that demonstrated that Saturn's moon Titan, exceptionally among the moons of the solar system, had an atmosphere.

The consideration of fixed-wing aircraft for Mars exploration has a long pedigree, with a central example being the delta-wing spaceplanes imagined by Wernher von Braun in Das MarsProjekt in 1952. The eventual determination of the extreme thinness of the Martian atmosphere by Mariner 4 in 1965, however, proved somewhat forbidding. Although the Viking landers used lifting entries (like Apollo), their landings were effected with rocket propulsion, and aeronautical ambitions‡ were held in abeyance.

After Viking, among the ideas kicked around for future Mars scientific exploration along with rovers and penetrators, were airplanes. A research project that helped solidify the notion of a Mars airplane was the Mini-Sniffer research aircraft developed by NASA's Dryden Research Center at Edwards Air Force base in California (see Fig. 1.6). This remotely piloted aircraft was intended to reach and sample the air at very high altitude (80,000 ft or more than 25 km). It had a pusher-propeller configuration so that the air samples acquired by equipment in the nose would be from clean air, free of engine exhaust or other contaminants. One version of the aircraft was developed to

Fig. 1.6 Two variants of the 1976 NASA Mini-Sniffer atmosphere-sampling, remotely piloted aircraft. The version on the right used a nonairbreathing engine using toxic hydrazine rocket fuel, hence the protective suits worn by the servicing crew.
Source: NASA Dryden images.

---

‡A rather obscure report "Vehicles for Exploration on Mars," published by the RAND Corporation in 1960, made brief calculations on a rocket-turbine–powered helicopter for carrying astronauts on Mars. However, the atmospheric density assumed for these calculations was an order of magnitude too high.

Fig. 1.7 A mockup of the VEGA balloon and gondola on display at the Udvar-Hazy Center of the Smithsonian Air and Space Museum. Notice the anemometer to the right of the gondola and the conical antenna at the top of the gondola.
Source: Author.

use hydrazine rocket fuel (hydrazine is a monopropellant that decomposes violently on a catalyst, without requiring an oxidant such as air in which to burn) that would also be suitable for operation on Mars.

However, the Mars airplane concepts remained immature, and the 1980s were an austere decade for planetary exploration in the United States. In fact, the only missions to reach another planetary surface in the 1980s were the Soviet VEGA-1 and VEGA-2 (see Fig. 1.7), which each dropped off a lander and a small helium balloon at Venus in 1985 on the way to Halley's comet. The VEGA balloons were the first extraterrestrial aerial platforms, and each made about 48 h worth of measurements near the cloudtops in Venus' atmosphere. (In the figure, the gondola and balloon are correct in size, but the tether connecting the two was much longer in flight. The long riser maximized the pendulum period and thereby minimized swing dynamical effects and the wake effect of the balloon on the instruments during ascent. The practical constraints on the display environment in the museum required a short riser.)

Seemingly out of nowhere, a conference paper in 1993 seemed to be the first technical study of an autonomous rotorcraft for Mars exploration (see Fig. 1.8).[4] In fact, I happened to see the talk myself at the International Astronautical Congress in Graz, Austria, where I had a talk as a student, and I recall that some aspects were a little hard to follow. The speaker was from

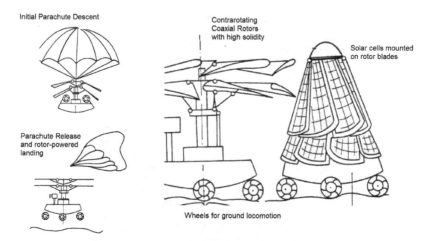

Fig. 1.8   Annotated sketch montage of elements of the cartoon-like concept by Savu et al. Although not an obviously practical design, the paper stimulated thinking about rotorcraft on Mars.
Source: Author.

(relatively) nearby Romania, a nation hardly at the forefront of planetary exploration, least of all in the turmoil just after the fall of the Iron Curtain, and thus the paper seemed purely academic.

At this time, a Mars balloon was under active development by the French and Russians for the Mars-96 mission. The balloon had to be the size of a bus merely to float a few kilograms of equipment. Unfortunately, the project saw delays and was eventually canceled.

There was growing interest in Martian aeronautics in the United States, as the gulf in scales of imaging possible from landers or rovers (centimeters to tens of centimeters in resolution, over a few tens to hundreds of meters) and orbiters (meters of resolution) became glaringly apparent. Balloon and airplane concepts were proposed to NASA programs (where they would have to compete with other mission ideas like lander networks or dust-skimming sample return craft), but without success.

For a short period, it seemed that a fixed-wing aircraft would fly on Mars. A competition was arranged to mark the centenary of the Wright brothers' flight in the 2003 Mars flight opportunity.[§] A substantial body of work was performed on some airplane concepts such as the Aerial Regional Environmental Survey (ARES) plane (see Fig. 1.9),[5] including a drop test from a high-altitude balloon, demonstrating deployment of its folding wings at 103,500 ft.

[§]The changing planetary alignments are such that practical opportunities to send spacecraft to Mars occur at roughly two-year intervals. Thus spacecraft flew to Mars in 1997, 1999, 2001 etc. In the end, the airplane did not fly, although the 2003 launch window would be used by the rovers Spirit and Opportunity, and the European Mars Express and Beagle 2 spacecraft.

Fig. 1.9   Full-size model of the ARES Mars airplane with its project team at the NASA Langley Research Center in 2003.
Source: NASA.

Fig. 1.10   Lightweight rotor being prepared for a test at Mars pressure in the Planetary Aeolian Facility at NASA Ames Research Center, circa 2000.
Source: NASA, courtesy L. Young.

Around the turn of the millennium, Larry Young, a rotorcraft engineer at NASA's Ames Research Center, began advocating for rotary-wing planetary exploration (see Fig. 1.10).[6] He outlined several concepts for Mars and also advocated vertical lift (ducted fan) approaches for Titan exploration.[7] In 2000, Ilan Kroo at Stanford University and the Jet Propulsion Laboratory (JPL)[¶] tested a small rotor under Mars-like atmospheric pressure in a JPL vacuum chamber.[8] The challenge of the thin Mars atmosphere, which demands high performance from any rotor, as discussed in the next chapter, is compounded by the poor aerodynamic behavior of most wing sections under Mars conditions (low Reynolds number). Kroo had led efforts to develop ultra-miniature rotorcraft (which he called a *mesicopter*) weighing just a few grams, and these small-scale vehicles confronted this same challenge. Kroo suggested that these tiny vehicles could be used in Mars or Titan exploration.

In 2000, Sikorsky Aircraft and NASA Ames Research Center cosponsored the American Helicopter Society's International Student Design Competition on the topic of a Mars rotorcraft. Entries were made by Georgia Tech (a quad-rotor[9]) and the University of Maryland. The latter entry (see Fig. 1.11), led by Anubhav Datta (who later became a professor at the same institution) was documented in somewhat more detail[10] and features many considerations that eventually became prominent during the development of Ingenuity.

Over the following years, a few sporadic publications documented experimental investigations or mission studies for Mars rotorcraft.[11]

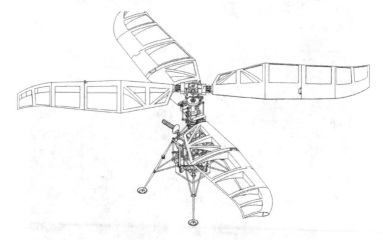

Fig. 1.11   The Mars autonomous rotary-wing vehicle (MARV) studied by the University of Maryland team led by Anubhav Datta for the 2000 AHS Student Design Competition. Source: Courtesy A. Datta.

[¶]The Jet Propulsion Laboratory in Pasadena, California, is a federally funded research and development center (FFRDC) owned and operated by the California Institute of Technology. Although JPL operates somewhat like a NASA center, it retains a distinct identity, and although subject to many of the same regulations, its employees are not government employees.

Meanwhile, the Mars exploration program proceeded apace, with the two Mars Exploration Rovers Spirit and Opportunity landing in 2004. The next landed Mars mission was competed, selected from a portfolio of proposed missions that included some with aerial elements; however, a (relatively) simple lander named Phoenix was selected to land in the ice-rich high northern plains.

The planetary community had slowly become socialized to the idea of planetary aerial exploration, thanks to the Mars airplane studies and ideas for Venus and Titan balloons. Now little quadcopters—just like Kroo's pioneering experiments—were beginning to appear in gadget stores in airports. Although the perennial challenges of weight and power remained, there was the sense that the problems of stability and autonomous control—essential issues because planetary distances mean that remote joystick flying by a human pilot would be impossible—had been solved.

In 2012, the large (1984-lb/900-kg) radioisotope-powered rover Curiosity (also known as the Mars Science Laboratory) landed at Gale Crater on Mars and began exploring. Its success, and programmatic considerations, prompted NASA to announce a similar follow-on rover that was scheduled to arrive in 2020. This vehicle would use a near-identical design to the $2.5B Curiosity, but would have a different suite of instruments to support documentation of samples to be returned to Earth. Perhaps this might be an opportunity for an add-on rotorcraft?

Some 44 instruments were proposed in response to the NASA solicitation for payloads for this Mars 2020 rover. These proposals were subjected to withering scrutiny by a review panel to assess not only the likely technical development risks and challenges of accommodation on the rover, but also the scientific value of the proposed payload and how it mapped to community-accepted scientific goals. Only seven payloads were selected for flight, and a helicopter (proposed by JPL Mars scientist Matt Golombek working with engineer Bob Balaram) was not among them. Soon thereafter, Balaram[12] and the team made their Mars helicopter concept public[13] in order to draw wider support for it, while some ongoing development was sustained on JPL internal funds.

A 2014 JPL progress report[14] indicated that the helicopter would have a mass of 1 kg (2.2 lbs). Individual daily flights would be limited to a short duration of approximately 3 min due to power, but should attain $\sim$100 m ($\sim$300ft) altitude and $\sim$600 m ground track (see Fig. 1.12). But to elevate a Mars helicopter from an internal research project would require some serious investment. JPL found support for its efforts in Senator John Culberson,[15] who was an advocate for space exploration and especially of the search for life on Jupiter's moon Europa.

Shown the prototype helicopter, Culberson worked to ensure that the 2017 House of Representatives budget provided $15 million to keep development of the helicopter on track to be ready for the 2020 rover mission. This critical funding enabled testing of a full-size vehicle in a vacuum chamber at Mars pressures, which in turn demonstrated sufficient progress to permit the inclusion of it on the 2020 rover.

Fig. 1.12   The Mars helicopter concept as envisioned in 2015. A distinctive feature of this early concept was the circular solar panel above the rotors. The practicalities of accommodation and fitting on the needed number of solar cells would later change this to a rectangle in the final design.
Source: NASA/JPL/Caltech.

On 11 May 2018, NASA announced that a Mars Helicopter Scout mission would fly.[16] "NASA has a proud history of firsts," said NASA Administrator Jim Bridenstine. "The idea of a helicopter flying the skies of another planet is thrilling. The Mars Helicopter holds much promise for our future science, discovery, and exploration missions to Mars." Culberson added:

> It's fitting that the United States of America is the first nation in history to fly the first heavier-than-air craft on another world. This exciting and visionary achievement will inspire young people all over the United States to become scientists and engineers, paving the way for even greater discoveries in the future.

Among those young people was 13-year-old Vaneeza Rupani of Northport, Alabama. When NASA ran a competition to give the helicopter a name, her proposal of Ingenuity was the winner. The 2020 rover itself had been named Perseverance in a similar competition.

When Perseverance reached Mars in 2021, it was announced that the Ingenuity helicopter represented a NASA investment of some $85 million.[17] The project has met with wide public interest and, because it has proven successful, apparent approval. Furthermore, the width of the Martian stage has grown, with the 2021 season seeing the arrival of not only Perseverance/Ingenuity, but also the United Arab Emirates' Hope orbiter and the Chinese Tianwen-1 mission, including the Zhurong-1 rover. The Ingenuity helicopter, then, served an arguably important political role in differentiating Perseverance from the Chinese efforts and maintaining a prominent perception of U.S. technological superiority.

At the time of writing, the helicopter has made more than a dozen flights and is still going strong.** Its technology demonstration objectives have been met, and indeed, far exceeded. Further, the hiatus in operations associated with the Mars Superior Conjunction (Mars being on exactly the other side of the sun, which impedes radio communications for a couple of weeks) makes a useful break.

Press statements by NASA have suggested the prospect of rotorcraft elements of many future Mars missions. Such statements are probably essential as post-hoc justification of the investment in technology development. Studies of larger and more scientifically capable rotorcraft for Mars have been published. Chapter 9 considers whether these are indeed realistic.

AIRCRAFT FOR TITAN

Saturn's moon, Titan, the only moon in the solar system with a significant atmosphere, seemed like a natural place to fly; even as far back as 1976, Titan mission concept studies included balloons.[18] But practical design of aircraft for Titan had to wait until the Voyager 1 spacecraft encounter in 1980, which determined the radius of Titan's solid surface (and thus its surface gravity) and the density of its atmosphere, both by a radio occultation experiment. Until that observation, there was an uncertainty of almost a factor of 100 in how thick the atmosphere might be.

Armed with that knowledge, a more detailed study of a Titan airship was made in 1984, but this seemed like a large step given the unknowns about Titan, whose surface had been hidden from Voyager's cameras by haze in the atmosphere. Instead, the Cassini mission—featuring a Saturn-orbiting spacecraft with a short-lived parachute-borne probe for Titan—was proposed in 1982. Scientific support for this mission grew in both Europe and the United States, and it was studied in progressively more detail, eventually being approved for flight in 1989/1990, with the probe to be named Huygens.[19]

In fact, a 1990 paper by Bob Zubrin (then at Martin Marietta) suggested the utility of tilt-rotor vehicles at Titan. These were intended as fetch vehicles to bring surface samples to a somewhat fantastical mother ship, a 10-t reactor-powered spacecraft. The utter incompatibility of this grandiose vision with the level of investment in the planetary exploration program of the time (or at any time!) notwithstanding, Zubrin's imaginative consideration of the ease of heavier-than-air flight at Titan was pioneering. Memorably, he wrote that a human could strap on a pair of wings and fly on Titan, a meme that the present author has used ever since in advocating Titan flight.

** As this book went through proofing in March 2022, Ingenuity had made more than 22 flights.

Fig. 1.13   In the early to mid-2000s, a hot air balloon or Montgolfière was advocated for Titan exploration. The vehicle would use buoyancy generated with the waste heat from a plutonium radioisotope generator.
Source: NASA image by artist Tibor Balint.

As the Cassini-Huygens mission reached the launch pad in 1997, thoughts began to turn to Titan exploration after Cassini. The starting point, technologically, remained lighter-than-air vehicles. Scientifically, it was recognized that Titan's surface would be compositionally diverse—the first surface maps had been made with data from the Hubble Space Telescope in 1994—and thus mobility would be an important capability for future exploration. Several studies of airships and balloons were performed at JPL in this period (see Fig. 1.13).[††]

This demanded consideration of the specific questions of where an uncontrolled balloon would go and how much power would an airship need to overcome winds and thereby access specific designated target areas. Another issue was how surface material might be accessed from a balloon or airship platform, prompting exploration of harpoons that might be retracted by a tether.

These challenges led the present author to realize that lighter-than-air vehicles might not be the right answer. Beyond the foregoing considerations is the fact that although the dense Titan atmosphere is favorable in terms of reducing the volume required for a buoyant envelope, the tankage required to bring the lifting gas (hydrogen or helium) is quite massive.

Whereas the dense atmosphere is favorable for lighter-than-air as well as heavier-than-air vehicles, only heavier-than-air vehicles draw the tremendous additional benefit that Titan's low gravity brings. This means that a rotorcraft

---

[††]I was involved in several of these studies. Ironically, one person I met with at JPL to discuss path planning and guidance for Titan airships in the mid-2000s was Bob Balaram.

of a given mass and rotor area on Titan would need only 1/38 of the power to hover that it would need on Earth. Even so, continuous flight would not be practical, given the modest power-to-weight ratio of the radioisotope power sources (RPSs) that would be needed so far from the sun.

However, Titan's slow rotation period meant that a vehicle at low or mid-latitude on Titan's surface would be out of view of the Earth (and in darkness) for eight days at a time, and during this local nighttime the continuous output of the RPS could be harvested and stored in a battery. The large energy store could then be used to support a short flight during the daytime, when the vehicle was in Earth view. This helicopter operations concept, published in 2000 (see Figs. 1.14 and 1.15), is essentially what became Dragonfly.[20]

At the time, unmanned aerial vehicles (UAVs) were for the most part remotely piloted vehicles with limited autonomy. The vast majority of these were fixed-wing, with only a couple of operational rotary-wing UAVs. The only prominent civilian example was the Yamaha R-MAX, a 44-kg (3 slugs, or 97 lbs) petrol-fueled helicopter used in Japan for crop-spraying and exploration of volcanos. In short, the technology was not ready to permit the realization of a Titan rotorcraft.

Fig. 1.14 An article on the Titan rotorcraft featured a particularly garish cover (and a rather impractical-looking vehicle design, somewhat resembling a Sikorsky S-37/ Westland Whirlwind).

Source: Author's collection.

Fig. 1.15 A helicopter inspecting the Huygens landing site, advocated by the author circa 2001. The coaxial rotors allow for relatively efficient packaging inside an entry shell, thus avoiding the need for a tail rotor. The concept was a couple of decades ahead of its time. Source: James Garry.

Thus, consideration of lighter-than-air concepts continued at JPL, although the airship had now fallen somewhat out of favor because of growing recognition of both the high power demands for propulsion[21] and the mass needed for lifting gas tanks. Instead, a hot air balloon (Montgolfière‡‡) came to the fore. Such a platform could exploit the waste heat from an RPS and would avoid the need for gas tanks. The idea of using waste heat in this way was aesthetically appealing, and the envelope of a hot air balloon would not be vulnerable to small punctures or leaks as would a light gas balloon. A Montgolfière offered the flexibility of control in the vertical dimension at least, with the prospect that perhaps some sort of horizontal control could be effected by exploiting altitude variations in winds, as performed by terrestrial balloon pilots.

In early 2005, after its release from Cassini, the Huygens probe made a 2.5-h parachute descent through the atmosphere of Titan, landing safely on the

---

‡‡Notice the grave accent on the è: the balloon type pioneered by the Montgolfier brothers is a Montgolfière. Their second balloon that made the famous 'first flight' from Versailles of a sheep, a duck and a rooster was named "Le Réveillon", after the owner of the paper mill. Keeping sponsors happy is an important element of innovation.

surface. This gave important ground truth on which to anchor the remote observations by Cassini, and the images from the probe turned Titan into a real place to explore. In 2007, NASA asked the Johns Hopkins University Applied Physics Lab (APL) to study a post-Cassini flagship mission, Titan Explorer. In considering the many options, heavier-than-air possibilities were quickly discarded as insufficiently mature; the adopted mission concept used an orbiter, a Montgolfière, and a lander, each of the three elements addressing different scientific goals. As for the Mars concepts in the previous 5–10 years, the aerial element offered regional survey capability, bridging the scale gap between landed and orbital observations.[§§] At this point, only a couple of years into Cassini's mission, Titan's seas had not been mapped in detail, and the largest area thought to be safe to land was in Titan's vast equatorial deserts, specifically the Belet sand sea. A lander with parachutes and airbags, like the Mars Pathfinder or Beagle, would simply roll to the base of a dune and unfurl its petals to begin science operations. The question naturally arose as to which element should be descoped if it were to prove unaffordable to fly all three. A rigorous assessment of the scientific value of the measurements to be made found that the lander (specifically, its contributions to surface chemistry and seismology) was much more valuable than the balloon. This result ran counter to the aesthetic appeal of flight, but was robust to different means of assessing science value. It was noted,[22] however, that an expendable, small (~1 kg) UAV could easily be launched from the lander to add to the appeal of aerial operations and regional imaging. A small vehicle would need to balance the heat generated by energy dissipation in its batteries against that lost to the cold environment in the same manner as subarctic insects, so the concept was named Titan Bumblebee.

Although the Titan Flagship study lost out to what became the Europa Clipper mission, it did serve to document the breadth of scientific opportunities on Titan and in particular what could be accomplished with a fixed lander and a balloon.

An exciting opportunity emerged in 2010 with the possibility of (just) squeezing a small Titan capsule into NASA's Discovery program, a series of roughly $500-million missions advocated by individual teams of scientists (rather than requiring endorsement by the community at large, via Decadal Surveys). The Titan Mare Explorer (TiME) mission would splash a capsule into Titan's second-largest (but best-mapped) polar sea of liquid methane, Ligeia Mare. The capsule—effectively a drifting buoy—would measure the composition of the sea and its depth with sonar and would study air–sea interactions. The team was led by planetary scientist Ellen Stofan (who later went on to become NASA's chief scientist) and included aerospace giant Lockheed Martin and APL. One reason such an ambitious mission could fit in

---

[§§]Ironically, this scale gap largely disappeared at Mars around this time, with the advent of the Mars Reconnaissance Orbiter's HiRISE camera with ~0.3-m resolution using a large telescope in a relatively low orbit. That gap-bridging would not occur with Titan, however, because the thick atmosphere demands a relatively high orbit (> 1000 km), and Titan's pervasive thick haze causes blurring by scattering.

the modest Discovery cost envelope was that a new power source, the Advanced Stirling Radioisotope Generator (ASRG), was to be provided by NASA. The other reason was that Ligeia would be in view of Earth when TiME arrived in 2023, Titan's northern midsummer. This meant that with a steerable antenna, TiME would be able to communicate directly with Earth without needing a relay spacecraft, which would be unaffordable.

Unfortunately, the development of the ASRG faltered, and the cost of pressing its development forward to be ready for TiME's launch in 2016 would be prohibitive, so in 2012 NASA selected instead the solar-powered InSight seismometer mission to Mars.

The prospect of a lightweight and efficient radioisotope power source had stimulated other Titan mission ideas. A team at NASA's Ames Research Center, with Titan scientist Jason Barnes, developed an airplane concept called the Aerial Vehicle for In-situ and Airborne Titan Reconnaissance (AVIATR).[23] In some ways this built on the previous Mars airplane work: the idea was that it would be in perpetual low-altitude flight, making high-resolution mapping and spectroscopic studies of the surface. Unfortunately, as the ASRG's predicted power output fell and its predicted mass grew (even so, it would be much more efficient than the MMRTG),¶ the concept proved not quite viable. The evolution of this project had interesting parallels in the energy/weight and power/weight challenges of early terrestrial aviation.

The next Discovery opportunity in 2015 specifically excluded missions requiring radioisotope power sources. This precluded any extended Titan surface mission concepts (including a reproposal of a TiME-like mission using an MMRTG), although there was a solar-powered spacecraft idea called Journey to Enceladus and Titan (JET) proposed by Christophe Sotin at JPL. An orbiter in free space can unfurl large panels without worrying about the atmospheric or gravity interactions that make solar power impossible on Titan's surface. Nonetheless, flying such large arrays (which need to be 100 times larger than those on Earth to generate the same power) is technically challenging, and JET was not selected in this Discovery call.

A NEW HOPE

The opportunity for what became Dragonfly may in some respects be seen as an unintended consequence of enthusiasm for Europa. Senator Culberson, keen to see a Europa lander mission,[24] chaired the Science Committee of the U.S. House of Representatives in 2015 and introduced a new element in NASA's budget, Ocean Worlds. The legislative language read as follows:

¶The MultiMission Radioisotope Thermoelectric Generator (MMRTG), used on Curiosity and Perseverance is an RTG variant that can operate in an atmosphere, unlike the RTGs used on Cassini, for example. The thermoelectric energy conversion of heat into electricity is about 4 times less efficient than a Stirling engine would be, but has the advantage of exceptional reliability and simplicity, since it is a solid-state system with no moving parts.

"The Committee directs NASA to create an Ocean World Exploration Program whose primary goal is to discover extant life on another world using a mix of Discovery, New Frontiers and flagship class missions consistent with the recommendations of current and future Planetary Decadal surveys,"[25] and called out Europa, Enceladus, and Titan as specific destinations.

In response to this directive, in early 2016 NASA indicated that it would solicit proposals for missions that explored Enceladus and Titan in its next New Frontiers solicitation. This naturally led to wide speculation that TiME would be reproposed. (What better fit for an Ocean Worlds program than a capsule that explored the seas on an icy moon?) But just as astrologers have always contended, our fates are at the mercy of planetary alignments. A mission that might start development in 2018–2019 would not launch until the mid-2020s and would arrive in the early 2030s, when it would be northern winter on Titan. Ligiea would be out of view, and direct-to-Earth communication would be impossible.

The New Frontiers budget envelope (stated to be $850 million, although bidders would have a good chunk of this deducted if they required an MMRTG as government-furnished equipment) was larger than that of Discovery, but even then a relay satellite would be prohibitively expensive. Thus, any Titan surface mission would have to go to low latitudes or the southern hemisphere.

In fact, it had become recognized that Titan's solid surface would be much more interesting chemically than its seas. Cassini data acquired in 2013–2017 showed that most of the northern seas and lakes were rather pure liquid methane—distilled like freshwater in Titan's hydrological cycle—rather than being some rich organic soup.

So thoughts turned back to the Titan Explorer study and the dune lander. As recognized by Carl Sagan well before the Cassini mission,[26] the most interesting surface materials from an origins-of-life standpoint would be sites where transient liquid water from an impact crater or cryovolcano might have interacted with the sand. Although all kinds of interesting missions and destinations could be devised for Titan, it seemed that to be responsive to the Ocean Worlds astrobiology emphasis, accessing this sort of material would be important.

These considerations were swirling around in Feb. 2016 when Jason Barnes and I were discussing ideas over dinner and he suggested a quadcopter (see Fig. 1.16).*** I immediately saw that although the MMRTG had been far too heavy for AVIATR, this didn't matter for a landed rotorcraft—an MMRTG would be a perfect fit! Furthermore, the charge-and-fly operations would be very resilient to any decline in power output in the Titan environment: You would be able to do the same operations, just take a little longer to recharge in

---

***In fact, this was not the first mention of a "Titan quadcopter". JPL's Larry Matthies had been funded in 2014 by NASA's Innovative Advanced Concepts (NIAC) program to study a Titan Aerial Daughtercraft, a small quadcopter, as a fetch rover or scout for a Titan balloon or lander. This concept was essentially identical to the Titan Bumblebee, which predated it by several years, although a novel element in Matthies's NIAC study was to emphasize the machine vision/navigation aspects that had not been considered in detail in the Bumblebee paper.

between flights. I quickly dusted off my Titan helicopter calculations with
their notional power source and recast them for an MMRTG-powered quad-
rotor, finding that in a single flight of an hour or so it would be possible to fly
further than any Mars rover had ever driven! Such mobility—now a much
more reasonable prospect, given the drone revolution, than it had been in 2000
when I had begun advocating for Titan rotorcraft—would let us scout for and
find the astrobiologically interesting materials.

Getting an institution to invest the million dollars or so of effort to assemble
a serious proposal to a NASA mission solicitation is no small matter, and
nothing as daring as this "relocatable lander"—basically picking up the whole
Curiosity-sized vehicle and flying it to multiple sites—had ever been
considered before. The details of how the team was assembled[27] and the
concept developed[28] are a story for another time, but suffice it to say here that
the combination of scientific appeal and that-just-might-work technical
achievability proved irresistible to APL. A small crack team, led by APL's Dr.
Elizabeth Turtle as principal investigator, submitted the proposal—named

**Fig. 1.16 A quadcopter drone circa 2018, one of the smallest rotorcraft available.[†††] The
widespread availability of hobby drones in the late 2010s did much to reduce the perceived risk of the Dragonfly proposal.**
Source: Author.

[†††]This drone belongs to the Dragonfly Project Manager, Peter Bedini. I flew it in idle moments in our
war room as we developed the Dragonfly concept.

Dragonfly—in April 2017. It was evaluated, along with 11 others (including Venus missions, a Saturn probe, and other concepts[29]), by NASA, which selected it in December of that year for a $4-million Phase A study, along with CAESAR, a comet sample return proposal. Over the following 15 months the concepts were developed further, with detailed designs, development plans, schedules, and budgets laid out for launch in 2025. After a grueling evaluation of both proposals, NASA chose Dragonfly as its next New Frontiers mission in June 2019.[30] "With the Dragonfly mission, NASA will once again do what no one else can do," said NASA Administrator Jim Bridenstine.[‡‡‡] "Visiting this mysterious ocean world could revolutionize what we know about life in the universe. This cutting-edge mission would have been unthinkable even just a few years ago, but we're now ready for Dragonfly's amazing flight."

NASA initially directed Dragonfly to plan for launch in 2026; later (as the COVID-19 pandemic introduced delays and costs across the agency) launch was moved to summer 2027. Dragonfly will arrive at Titan by 2034, almost exactly one Titan year after Huygens (an important coincidence, in that the season will be the same so Huygens's measurements of the atmospheric properties are directly applicable).

Because Dragonfly is still in development, some of the details outlined in Chapters 7 and 8 may see modest changes. Nonetheless, it is now five years since the concept was originated, so the design has largely converged (see Fig. 1.17), making now a good time to report on it.

CODA

It is interesting to reflect that planetary aeronautics now is very much where terrestrial aeronautics was almost exactly one century ago (Table 1.1). The competing technologies of fixed- and rotary-wing, as well as lighter-than-air, flight are slowly shaking out to find their respective niche applications. The technical hurdles of power-to-weight and control are the same as they have always been but are compounded by the additional challenges of planetary operation including thermal control, remote operations with large latency, and the rigors of delivery to a planetary surface.

Although the public not unreasonably assumes a direct connection between Ingenuity and Dragonfly, and indeed NASA itself sometimes reinforces this notion, these two projects have very different—and quite independent— programmatic origins. One is an uncompeted 2-kg technology demonstration with the sole objective of demonstrating flight, that was added on to an existing planetary rover mission that provides delivery and communications services,

---

[‡‡‡]Historically, such modest affairs as technology demonstrators or even science missions would be released by officials lower in the NASA organization. However, Administrator Bridenstine, himself a former Navy pilot, evidently had enough enthusiasm for planetary aeronautics to participate in the announcement himself.

Fig. 1.17 Dragonfly mission concept. After delivery from space in an aeroshell and parachute, the vehicle lands under rotor power and deploys a high-gain antenna for direct communication with Earth. Powered by a radioisotope power supply that provides heat and trickle-charges a large battery, the vehicle can operate nearly indefinitely as a conventional lander, but can make periodic, brief, battery-powered rotor flights to new locations. Source: APL.

TABLE 1.1 TIMELINE OF PLANETARY FLIGHT

1976	Martin Marietta study suggests airships for Titan.
1984	Soviet VEGA balloons launch to Venus.
1993	Savu describes Mars helicopter concept.
1996	Intended launch of Russian/French Mars-96 balloon (canceled).
2000	Lorenz advocates hibernating helicopter concept for Titan. Young advocates rotorcraft generally.
2000	Kroo explores small rotorcraft concepts.
2003	Dutta performs Mars helicopter design study.
2007	Titan Explorer Flagship study suggests Titan Bumblebee daughtercraft.
2012	Titan Mare Explorer evaluated, but not selected. Mars 2020 rover announced.
2014	JPL discloses Mars helicopter work.
2016	Dragonfly conceived (January). Mars helicopter makes first flight.
2017	Dragonfly proposed to New Frontiers 5 solicitation. Selected for $4-million Phase A study (December).
2018	Mars Helicopter Scout implementation announced.
2019	Dragonfly selected for flight.
2020	Mars Helicopter Scout (now Ingenuity) and Perseverance rover launch.
2021	Perseverance lands on Mars. Ingenuity performs powered flight on Mars.
2027	(Planned) Launch of Dragonfly.
2033/4	(Planned) Arrival of Dragonfly at Titan.
2037	(Planned) End of Dragonfly nominal mission.

whereas the other is a more-than 800-kg self-contained exploration platform that happens to be a rotorcraft because that is the best way to achieve its peer-reviewed scientific objectives.

Regardless of these different genetic circumstances, both projects are triumphs of innovation that realize humankind's dream of flight but on other worlds. They are therefore of great public interest, and it is the author's hope that this book can help explain how these vehicles work.

REFERENCES

1 There are many good histories of helicopter development. The chapter on this topic in Gordon Leishman's text *Aerodynamics of the Helicopter* (Cambridge, 2000) is recommended. The compilation by E. Liberatore, *Helicopters Before Helicopters* (Krieger, 1998) is fascinating, although largely limited to developments in the United States. An invaluable resource is the three-volume, 2000+-page epic history, *Introduction to Autogyros, Helicopters, and Other V/STOL Aircraft*, published between 2011 and 2016. A NASA publication, *Introduction to Autogyros, Helicopters, and Other V/STOL Aircraft* (NASA/SP–2011-215959), is free for download, currently at https://rotorcraft.arc.nasa.gov. NASA publications in general can also be found on https://ntrs.nasa.gov.

2 Lorenz, R. D., "Edmond Halley's Aeronautical Calculations on the Feasibility of Manned Flight in 1691," *Journal of Aeronautical History*, Paper 2012/02.

3 The story of how we have attempted to measure and understand atmospheric conditions on other worlds is told in my book, *Exploring Planetary Climate: A History of Scientific Discovery at Earth, Mars, Venus and Titan* (Cambridge, 2019).

4 Savu, G., Trifu, O., and Oprisiu, C., "An Autonomous Flying Robot for Mars Exploration," 44th Congress of the International Astronautical Federation, IAF Rept. 93-U.4.568, Oct. 1993. See also Savu, G., and Trifu, O., "Photovoltaic Rotorcraft for Mars Missions," *Proceedings of the Joint Propulsion Conference and Exhibit*, San Diego, CA, 1995.

5 Braun, R. D., Wright, H. S., Croom, M. A., Levine, J. S., and Spencer, D. A., "Design of the ARES Mars Airplane and Mission Architecture," *Journal of Spacecraft and Rockets*, Vol. 43, No. 5, 2006, pp. 1026–1034.

6 Among Young's presentations and publications are Young, L. A., "Vertical Lift—Not Just for Terrestrial Flight," *Proceedings of the AHS/AIAA/RaeS/SAE International Powered Lift Conference*, Arlington, VA, 2000; Young, L. A., Chen, R. T. N., Aiken, E. W., and Briggs, G. A., "Design Opportunities and Challenges in the Development of Vertical Lift Planetary Aerial Vehicles," *Proceedings of the American Helicopter Society International Vertical Lift Aircraft Design Conference*, San Francisco, CA, 2000; Young, L. A., Aiken, E. W., Gulick, V., Mancinelli, R., and Briggs, G. A., "Rotorcraft as Mars Scouts," *Proceedings of the IEEE Aerospace Conference*, Big Sky, MT, 2002; Young, L. A., Lee, P., Briggs, G., and Aiken, E., "Mars Rotorcraft: Possibilities, Limitations, and Implications for Human/ Robotic Exploration," *Proceedings of the IEEE Aerospace Conference*, Big Sky, MT, 2005.

7 Young, L.A., "Exploration of Titan Using Vertical Lift Aerial Vehicles," *Forum on Innovative Approaches to Outer Planetary Exploration 2001–2020*, Lunar and Planetary Institute, Feb. 2001, http://www.lpi.usra.edu/meetings/outerplanets2001/. I made my own early advocacy for Titan rotorcraft, "Flexibility for Titan Exploration—The Titan Helicopter" at the same meeting.

8 Kroo, I., and Kunz, P., "Development of the Mesicopter: A Miniature Autonomous Rotorcraft," *Proceedings of the American Helicopter Society Vertical Lift Aircraft Design Conference*, San Francisco, CA, 2000. See also Kroo's report to the NASA Institute for Advanced Concepts (NIAC), "The Mesicopter: A Miniature Rotorcraft Concept Phase II Final Report," http://www.niac.usra.edu/files/studies/final_report/377Kroo.pdf.

9   O'Brien, P., "Using a Robotic Helicopter to Fuel Interest in and Augment the Human Exploration of the Planet Mars," *AIAA Space 2003 Conference & Exposition*, p. 6275.

10 Datta, A., Roget, B., Griffiths, D., Pugliese, G., Sitaraman, J., Bao, J., Liu, L., and Gamard, O., "Design of a Martian Autonomous Rotary-Wing Vehicle," *Journal of Aircraft*, Vol. 40, No. 3, 2003, pp. 461–472. See also Datta, A., Roget, B., Griffiths, D., Pugliese, G., Sitaraman, J., Bao, J., Liu, L., and Gamard, O., "Design of the Martian Autonomous Rotary-Wing Vehicle," *Proceedings of the AHS Specialist Meeting on Aerodynamics, Acoustics, and Test and Evaluation*, San Francisco, CA, 2002. I met Anubhav Datta when APL brought him in for a "red team" review of the Dragonfly concept in 2018.

11 Young, L. A., Aiken, E. W., Derby, M. R., Demblewski, R., and Navarrete, J., "Experimental Investigation and Demonstration of Rotary-Wing Technologies for Flight in the Atmosphere of Mars," Proceedings of the American Helicopter Society Annual Forum, Montreal, Canada, 2002. Also Tsuzuki, N., Sato, S., and Abe, T., "Conceptual Design and Feasibility for a Miniature Mars Exploration Rotorcraft," *Proceedings of the International Congress of the Aeronautical Sciences*, 2004, and Song, H., and Underwood, C., "A Mars VTOL Aerobot—Preliminary Design, Dynamics and Control," *Proceedings of the IEEE Aerospace Conference*, 2007.

12 Balaram had seen a presentation by Kroo in the late 1990s. United Press International, "After Intense Testing, Mars Helicopter Ingenuity Ready to Fly," *Gephart Daily*, https://gephardtdaily.com/national-international/after-intense-testing-mars-helicopter-ingenuity-ready-to-fly/. Balaram told the story at the IEEE Aerospace Conference in Big Sky, MT in March 2022 that he had proposed some technology development work with Kroo in the early 2000s and it was well-reviewed but NASA was unable to fund it at the time. Someone recalled this proposed work to the JPL director in 2012, leading to his approaching Balaram to initiate the Mars Helicopter Scout effort that became Ingenuity.

13 Balaram, J., and Tokumaru, P. T., "Rotorcrafts for Mars Exploration," *Proceedings of the International Planetary Probe Workshop*, Pasadena, CA, 2014. LPI Contribution No. 1795.

14 Volpe, R., "2014 Robotics Activities at JPL," *International Symposium on Artificial Intelligence, Robotics and Automation in Space (i-SAIRAS)*, Montreal, Canada, 17 June 2014, https://www-robotics.jpl.nasa.gov/publications [retrieved 2 Sept. 2019].

15 Berger, E., "Four Wild Technologies Lawmakers Want NASA to Pursue," *Ars Technica*, 24 May 2016, https://arstechnica.com/science/2016/05/four-wild-technologies-lawmakers-want-nasa-to-pursue/

16 NASA, "Mars Helicopter to Fly on NASA's Next Red Planet Rover Mission," NASA Release 18-035, 11 May 2018, https://www.nasa.gov/press-release/mars-helicopter-to-fly-on-nasa-s-next-red-planet-rover-mission

17 The inclusion of the helicopter did not meet with universal approval; see, for example, Rafkin, S., "The Mars Ingenuity Helicopter: An Aerial Assault on the Integrity of Science and the Mission Peer Review Process," 13 April 2021, http://freethoughtranch.blogspot.com.

18 See, e.g., Lorenz, R. D., "A Review of Titan Mission Studies," *Journal of the British Interplanetary Society*, Vol. 62, 2009, pp. 162–174.

19 I was immensely fortunate to be assigned to work on the just-beginning Huygens project at the European Space Agency in Noordwijk, The Netherlands in October 1990 in my first job, right out of university. The story and details of Huygens's development and Titan exploration in the lead up to and during the Cassini mission are told in my books *Lifting Titan's Veil* (Cambridge, 2002), *Titan Unveiled* (Princeton, 2008/2010), and *NASA/ESA/ASI Cassini-Huygens Owners' Workshop Manual* (Haynes, 2017).

20 My paper, "Post-Cassini Exploration of Titan: Science Rationale and Mission Concepts," *Journal of the British Interplanetary Society*, Vol. 53, 2000, pp. 218–234, underscores the

importance of the energy source (note that the MMRTG had not yet been developed), notes the advantages of a rotorcraft, and outlines the charge-up-then-fly concept of operations and its relation to the Titan diurnal cycle. The popular article on the helicopter that followed the paper is Lorenz, R. D., "Titan Here We Come," *New Scientist*, 15 July 2000, pp. 24–27—it concludes "now I must get other scientists interested and get NASA to perform a detailed technical study." I would have been horrified to think it would take 17 years to do so! On the other hand, I did at least recognize "it's too early to know what this machine will actually look like." - indeed Dragonfly looks nothing like Fig. 1.15.

21 Lorenz, R. D., "Scaling Laws for Flight Power of Airships, Airplanes and Helicopters: Application to Planetary Exploration," *Journal of Aircraft*, Vol. 38, 2001, pp. 208–214. A small NASA Langley study several years later considered a Titan helicopter using an entirely hypothetical radioisotope-powered Brayton cycle engine, see R. Prakash, R. Braun; L. Colby; S. Francis; M. Gunduz; K. Flaherty; J. Lafleur; H. Wright, "Design of a long endurance Titan VTOL vehicle," 2006 IEEE Aerospace Conference, 2006, doi: 10.1109/AERO.2006.1655732. The helicopter did not fit into the parameters of the overall study, which instead selected an airship - see J. Levine and H. Wright "Titan Explorer" available at https://ntrs.nasa.gov

22 Lorenz, R. D., "Titan Bumblebee: A 1-Kg Lander-Launched UAV Concept," *Journal of the British Interplanetary Society*, Vol. 61, 2008, pp. 118–124.

23 Barnes, J., L. Lemke, R. Foch, C. P. McKay, R. A. Beyer, J. Radebaugh, D. Atkinson, R. Lorenz, S. Le Mouélic, S. Rodriguez, J. Gundlach, F. Giannini, S. Bain, F. M. Flasar, T. Hurford, C. M. Anderson, J. Merrison, M. Ádámkovics, S. Kattenhorn, J. Mitchell, D. M. Burr, A. Colaprete, E. Schaller, A. J. Friedson, K. S. Edgett, A. Coradini, A. Adriani, K. M. Sayanagi, M. Malaska, D. Morabito and K. Reh, "AVIATR—Aerial Vehicle for In-Situ and Airborne Titan Reconnaissance," *Experimental Astronomy*, Vol. 33, 2012, pp. 55–127, https://doi.org/10.1007/s10686-011-9275-9

24 The influence of Culberson and the story of the emergence of the Europa mission is engagingly told in David Brown's book, *The Mission: A True Story* (HarperCollins, 2021).

25 FY2016 House Appropriations Report 114–130 p. 59, U.S. House of Representatives.

26 Thompson, W. R., and Sagan, C., "Organic Chemistry on Titan-Surface Interactions," *Symposium on Titan*, ESA SP-338, European Space Agency, Noordwijk, The Netherlands, 1992, pp. 167–176.

27 The more general problem of forming and evaluating mission teams and concepts is described in my piece "The Unnatural Selection of Planetary Missions," https://www.thespacereview.com/article/2808/1

28 Lorenz, R. D., Turtle, E. P., Barnes, J. W., Trainer, M. G., Adams, D. S., Hibbard, K. E., Sheldon, C. Z., Zacny, K., Peplowski, P. N., Lawrence, D. J., Ravine, M. A., McGee, T. G., Sotzen, K. S., MacKenzie, S. M., Langelaan, J. W., Schmitz, S., Wolfarth, L. S., and Bedini, P., "Dragonfly: A Rotorcraft Lander Concept for Scientific Exploration at Titan," *Johns Hopkins Technical Digest*, Vol. 34, No. 3, 2018, pp. 374–387. See https://dragonfly.jhuapl.edu/News-and-Resources/

29 Kane, V., "Selecting the Next New Frontiers Mission," *Planetary Society*, https://www.planetary.org/articles/20160829-selecting-the-next-new-frontiers-mission

30 NASA, "NASA's Dragonfly Will Fly Around Titan Looking for Origins, Signs of Life," Release 19-052, https://www.nasa.gov/press-release/nasas-dragonfly-will-fly-around-titan-looking-for-origins-signs-of-life/

Chapter 2

BASICS OF ROTORCRAFT FLIGHT

"A helicopter doesn't fly; it beats the air into submission."

—Author unknown

This brief chapter should not be considered a substitute for the excellent texts that exist on helicopter aerodynamics and design.[1] However, it attempts to lay out the essentials to explain what the limits on rotor flight might be in different environments. I have somewhat deliberately presented the material in a manner distinct from the traditional helicopter conventions. Readers who are established in those approaches may not need to learn much from this chapter anyway, whereas those who are new to rotorcraft may find this perspective more accessible, or at least complementary to them. In the interests of readability, equations have been banished to the Appendix.

ATMOSPHERES

The principal atmospheric parameters of interest in aerodynamics are air density and viscosity. Density is by far the more important of these. On Earth at sea level, air has a density of 1.25 kg/m^3 (0.0023 slugs/ft^3). (Your lungs hold about 5 g of air.) Planetary surface environments of interest span more than two orders of magnitude in air density—the density on Titan is a factor of 4 higher than on Earth, but at most of the Martian surface, the atmospheric density is about 60 times less than Earth's. In fact, the density of the Martian atmosphere is highly variable—at a fixed location, the density and pressure vary by about 30% over the course of the year because the dominant atmospheric constituent, carbon dioxide, freezes out in the polar regions as a seasonal frost. Furthermore, the Martian surface spans a rather wide altitude range; the northern hemisphere is low-lying and relatively flat (likely a former seabed), and the south is a rugged highland. The irregular boundary between

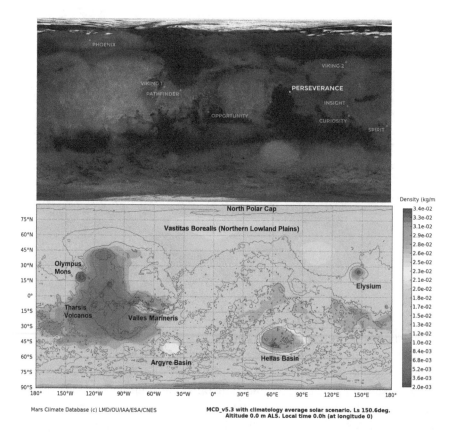

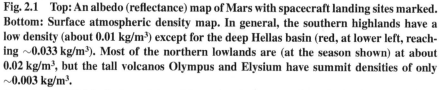

Fig. 2.1 Top: An albedo (reflectance) map of Mars with spacecraft landing sites marked. Bottom: Surface atmospheric density map. In general, the southern highlands have a low density (about 0.01 kg/m³) except for the deep Hellas basin (red, at lower left, reaching ~0.033 kg/m³). Most of the northern lowlands are (at the season shown) at about 0.02 kg/m³, but the tall volcanos Olympus and Elysium have summit densities of only ~0.003 kg/m³.

Source: Top: NASA; Bottom: Adapted by author from output from the LMD Mars Climate Database (http://www-mars.lmd.jussieu.fr/mars/access.html).

the two is punctuated by craters, canyons, and large volcanos. Moving from the base of the Tharsis volcanos, for example, where the pressure is about 6 mbar (0.087 pounds/inch²) to the summit is an elevation change of around 22 km (~14 miles), and the pressure and density drop by almost a factor of 6! Because the mechanics of entry and descent (discussed in the next chapter) are easier for larger surface pressures, almost all Mars missions have been aimed at the northern lowlands (see Fig. 2.1).

Note that in most of this book we consider only Mars and Titan. Although the atmosphere on Venus has an appealingly high density at the surface (50 times higher than Earth), it is accompanied by temperatures (750 K) (890 degrees Fahrenheit) that are intolerable for sustained operations of current engineering systems. Temperatures are benign at an altitude of around 60 km, where the atmosphere (which, like Mars, consists of mostly CO_2) also has a similar density to Earth's. Indeed, this is the region explored by the VEGA balloons in 1985; however, being so far from the surface, the benefits of a rotorcraft are not clear. We will return to this question at the end of the book in Chapter 9. The same arguments apply for the relevant levels in the atmospheres of the giant planets Jupiter and Saturn, and the ice giants Uranus and Neptune.

The atmospheric pressure is an indication of the total amount of gas in the atmosphere. It is the weight of the column of gas, and hence has dimensions of force per unit area. Indeed, we used to determine our atmospheric pressure by balancing it against the weight of a column of a different fluid—mercury. The weight of a $\sim$10-km column of air with a density of 1.25 kg/m^3 is the same as a 760-mm column of mercury with a density of 13,600 kg/m^3 (or, of interest to divers, a 10-m column of water with a density of 1000 kg/m^3). These are equivalent to one atmosphere, or the typical terrestrial sea-level pressure of 1.013 bar (1013 mbar). The formal SI unit of pressure is the Pascal (Pa), and 1 bar = 100,000 Pa (or 1000 hPa) or, in Renaissance units, 14.6 pounds per square inch.

Density, pressure, and temperature are related. For Earth and Mars, the relationships are the simple ideal gas laws, density being proportional to pressure and inversely to temperature. The most challenging conditions for aircraft are "hot and high," when the density is low. At the summit of Mauna Kea on Hawaii (a familiar location to astronomers, with an altitude of 4200 m) or almost 14,000 ft where pressure $P = 611$ mbar (8.8 pounds per square inch), we find the density to be about 0.8 kg/m^3, or one-third less than at sea level, which accounts (via the corresponding low partial pressure of oxygen) for the symptoms of altitude sickness often encountered there. The world helicopter altitude record is about 13 km, where the density was about 0.3 kg/m^3 (see Fig 2.2).

On Titan, the atmosphere is thick and cold enough to be somewhat close to partial condensation. The air on Titan is only 100 times denser than the methane–nitrogen liquid that forms Titan's seas, and the ideal gas laws are off by several percent; more complex equations of state need to be used for accurate calculations.

The 10-km column mentioned previously is the scale height H of the atmosphere. Unlike liquids, gases tend to be compressible, so the density declines gradually with height and there is no sudden top of the atmosphere. The decline is usually an exponential function of height, and H is the height distance over which the density falls by a factor of $e = 2.718....$ On the terrestrial planets, the value of H is 8–10 km, although for Titan with its low gravity, $H = \sim$21 km.

Two other properties of the atmosphere are of particular aeronautical interest. One is the speed of sound, the velocity at which a perturbation in pressure propagates through the air. The ratio of the flight speed (or the blade tip speed) to the speed of sound is called the Mach number M. Subsonic ($M < 1$) and supersonic ($M > 1$) flows have quite different properties.

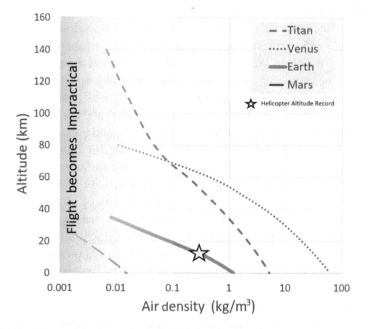

Fig. 2.2 Density vs altitude on the terrestrial planets and Titan. Aerodynamic forces become weaker at lower densities, and a value of ~0.01 kg/m³ denotes a more or less practical limit, which the Martian surface just barely exceeds. This threshold, more or less where the Huygens probe's parachute deployed, was at an altitude of ~150 km on Titan due to the large-scale height on that world.
Source: Author.

The other static property of a gas that is of interest is viscosity μ, which is a measure of how well momentum diffuses through a fluid. The dynamic viscosity[*] is the ratio of shear stress to strain rate (or velocity gradient) in the fluid and has SI units of Pa-s. It is a function of both composition and temperature. In air at room temperature, $\mu = 18\ \mu$Pa-s, whereas on Titan (also mostly nitrogen, but rather colder) the viscosity is almost 3 times lower. On the other hand, carbon dioxide has a slightly higher viscosity than nitrogen.

The ratio of viscous forces to inertial forces (i.e., those caused by the fluid's mass density) is called the *Reynolds number*, and it is another important defining parameter of the flow regime. At very low Reynolds numbers, the viscous forces dominate and flow is often very smooth or laminar, typically following the contours of the vehicle. As the Reynolds number increases, the flow may break away, or separate, from the blade or vehicle surface, changing the lift and drag characteristics. At high Reynolds numbers the flow is turbulent.

---

[*]Confusion sometimes arises because in some fields it is common to refer to kinematic viscosity, which is just dynamic viscosity divided by density. Physically this is a momentum diffusivity, and it has the same units (m²/s) as, for example, thermal diffusivity. Generally, if unspecified, then dynamic viscosity is meant; however, to be safe one should state explicitly which is intended.

In evaluating a new aircraft design, it is sometimes impossible and generally difficult and/or expensive to build a full-scale test article and test it with all the flight conditions replicated. However, to have the flow behave correctly (and thus to estimate parameters of interest such as lift and drag), it is only necessary to match the dimensionless parameters, Mach and Reynolds number. This allows useful insights from small-scale models, which are much cheaper to build and test. The test fluid does not even need to be air—useful insights can be gained from water tunnels, as long as the Reynolds number is being matched.[†] The Reynolds numbers of various planetary atmospheres are shown in Table 2.1.

A final property that occasionally is important is the mean free path, the distance an air molecule may statistically travel before it hits another air molecule. Generally this distance is very short. (In sea-level air on Earth it is about 70 nm, much smaller than any object we are considering.) The distance increases at lower pressures (densities) and can be several microns at Mars surface conditions, which is comparable with the size of fine dust particles and fibers. Thus at Mars surface conditions, the flow through dust filters, or the performance of thermal insulation, may need to take mean free path effects into account.[‡]

Table 2.1   Properties of Planetary Atmospheres

Body	Earth	Mars	Titan
Planetary radius (km)	6370	4470	2575
Gravitational acceleration (m/s^2)	9.8	3.7	1.35
Diurnal period (solar days)	1	1.04	16
Annual period (years)	1	1.88	29.5
Surface atmospheric pressure (bar)	1	0.006	1.46
Surface atmospheric temperature (K)	250–310	150–280	92–95
Dominant gases	N_2, O_2 (21%)	CO_2	N_2, CH_4 (5%)*
Atmospheric density (kg/m^3)	1.25	0.015 (typ)	5.4
Speed of sound (m/s)	340	~250	195
Dynamic viscosity (10^{-6} Pa-s)	17	13	6
Planetary boundary layer (km)	0.3–3	1–10+	0.3–1[†]

*Like water vapor on Earth, the methane concentration on Titan varies (eg, with altitude).
[†]The 0.3-1km is the classic diurnal boundary layer. In fact the 3km inflection in the atmosphere temperature profile may be somewhat different in origin, but has a similar capping effect (see text).

[†]An important Reynolds number regime is 10,000–100,000, where (depending on surface roughness) the transition of a flow from laminar to turbulent takes place. For reference, the Mars helicopter rotor blades operate at a Reynolds number ~11,000; those of a typical terrestrial helicopter or Dragonfly are at a few million.

[‡]The ratio of mean free path to object size is called the Knudsen number. In very rarified air, at orbital altitudes, the mean free path becomes comparable with the size of a spacecraft, and the drag coefficient of any shape approaches 2.0. Essentially, the air molecules become independent, noninteracting billiard-ball particles and the flow regime is called *free-molecular* or *Newtonian*. Isaac Newton imagined this was how aerodynamics worked; he was wrong, except in this near-space regime.

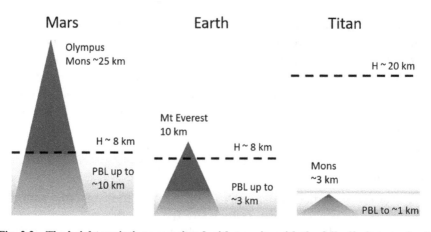

Fig. 2.3   The height variations associated with terrain, with the fall-off of atmospheric density H and the PBL depth on different worlds. Martian terrain has a large topographic variation compared with the scale height, and so surface pressure is very location-dependent.
Source: Author.

THE PLANETARY BOUNDARY LAYER

A distinct feature (Fig. 2.3) in the vertical structure of a world's atmosphere is the planetary boundary layer (PBL). The PBL is the layer that is well-mixed by convection from the ground; therefore, it has a uniform profile of certain thermodynamic properties, notably the potential temperature. The PBL typically grows in thickness throughout the day, as solar heating stirs the air near the surface. Over the oceans on Earth, where the heat capacity of the sea is large, the PBL is typically a few hundred meters thick. Over the continents, and in large deserts especially, the PBL can grow to be 2- to 3-km (1-2 miles) deep in the late afternoon. Sometimes the PBL is capped by an appreciable temperature inversion and may be visible because the air in the PBL is smoggy.

The PBL is of interest for a number of reasons. Because the PBL behaves in some ways distinctly from the rest of the atmosphere, the relationship of its thickness to the size of topographic obstacles affects how the air flows around and/or over them. Similarly, the size of sand dunes can be influenced by the PBL depth; indeed, it is possible that the remarkable uniformity of the size of Titan's sand dunes may be because they are controlled by that parameter.[2] Meteorological features can also be controlled by the PBL depth—notably, the size of dust devils seems to scale with it, and so dust devils are rather larger on Mars than on Earth. The dominant length scale of turbulent variations in wind also tends to be around one-third of the PBL depth.

On Titan, the PBL (controlled in part by the length of the day, the solar heating, and the heat capacity of the land and atmosphere) is believed to grow from about 300 m (as measured by the Huygens probe when it descended at around 0945 local solar time) to about 900 m in the late afternoon. However, the Huygens profile also showed residual structures—one perhaps the vestige of the previous day's PBL—at about 1 km, and other structures at 2–3 km. These latter features may be seasonal in nature or perhaps controlled by the distribution of greenhouse warming in the lower atmosphere. Whatever their origin, they may be responsible for the ∼3 km dune spacing seen all over Titan. This PBL structure is of relevance to rotorcraft design, in that the presence of these features in the profile is of great meteorological interest; therefore, the Dragonfly mission is being designed to be able to attain these altitudes to investigate the PBL structure. The dense atmosphere on Titan barely responds to the day/night cycle, despite its length, and the day/night temperature variation is only about 1 K, or 1%.

On Mars, the thin atmosphere needs very little heat to be warmed appreciably, and so the PBL can grow to be quite deep—as much as 10 km or more. Similarly, there is a very wide swing in air temperature from day to night (often around 80 K, or about 30%; see Fig. 2.4), and calculations for aerodynamic performance must bear the wide range in atmospheric properties accordingly.

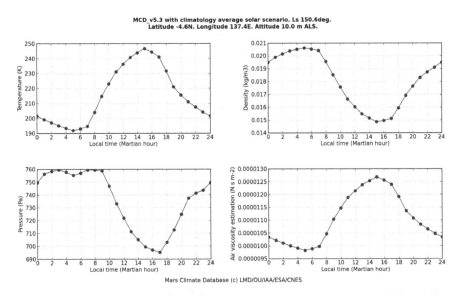

Fig. 2.4   The diurnal cycle of near-surface atmospheric properties on Mars (for the Curiosity rover landing site). Of note are the wide swings in pressure (almost 10%) and temperature (around 70 K). These result in a change of some 30% in air density over the course of the day, making some times of day much easier for flight than others.
Source: Author, using Mars Climate Database.

ROTARY WING FLIGHT

A rotor[§] is in essence a wing. It develops thrust by propelling air downward, and by conservation of momentum the thrust equals the momentum flux imparted to the air. This momentum flux is simply the mass flux times the increase in velocity of the air stream, so a given amount of thrust can be developed by hurling down some material at very high velocity or a much larger volume of material at lower velocity. The choice between these directions depends on the application—thrust in high speed flight requires a high exhaust velocity, whereas energy efficiency is maximized for low velocity.

A useful idealization of a rotor is the *actuator disk*—basically imagining that all the clever twists and camber of the spinning rotor blades manifests simply in a "magic" disc that accelerates air through it. Some simple but powerful insights, not only for hover but also for forward flight, result from considering the momentum balance in such a picture (see Fig. 2.5).

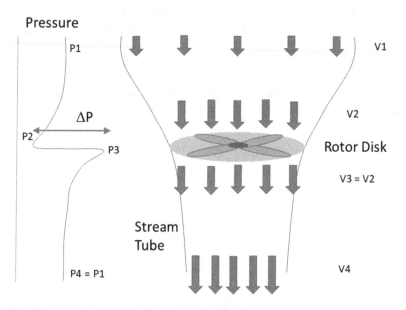

Fig. 2.5 The actuator disk idealization posits that the net effect of a rotor is to push a uniform circular column of air downward. The disk thrust equals the rate at which momentum is added to the air, which also equals the disk area times the pressure differential across it ΔP. In reality, the swirling flowfield is much more complex and nonuniform.
Source: Author.

[§]In this book I unapologetically use the word *rotor* in a generic manner. The reader should be aware that some in the rotorcraft industry insist the term *rotor* should be reserved for those whirly bits that have cyclic pitch control; whirly bits that only develop thrust are mere "propellers." A distinction between *propeller* and *fan* based on disc solidity has even been suggested. (See *Vertiflite* magazine, Nov/Dec. 2021). Like those that surround the definition of *planet*, nomenclature debates are always passionate, but rarely useful.

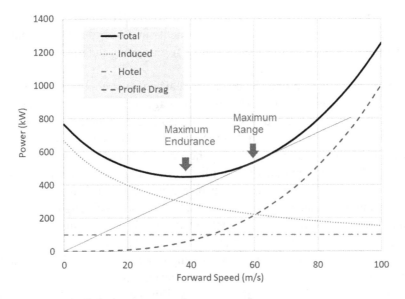

Fig. 2.6   A generic flight power vs speed curve.
Source: Author.

For example, in forward flight with the rotor axis tilted slightly to provide a forward component of thrust to balance any drag, air is rammed into the rotor disk. This leads to a reduction in the power needed to develop a given amount of thrust and explains why helicopters seeking to maximize their endurance (eg, for surveillance tasks, observing traffic, etc.) tend to fly in a circle around their target of interest rather than hovering over it.

The improved efficiency from forcing air into the rotor disk can only get you so far. As forward speed is increased, the drag on the helicopter fuselage increases (as the square of the speed), and eventually becomes the dominant drag term. The combination of the decline of the (thrust) induced power with speed and the cubic growth of the profile power term leads to a power vs flight speed curve that has a basin. At this curve's lowest point—the minimum power speed—a given amount of energy or fuel will last longest; this is the maximum endurance speed (see Fig. 2.6). In the figure, the hotel loads are background power needs for equipment and are invariant with speed. The induced power is that required to generate lift by the rotor—in forward flight this is actually less than in hover because air is rammed into the rotor disk. However, the profile drag term becomes dominant at high speed. The combination of these terms leads to a total power curve with a minimum value (minimum power = maximum endurance), which is at a lower speed than the maximum range speed, which is defined by minimum power divided by speed.

There is a slightly higher speed at which a line from the origin forms a tangent to the curve. This is at a slightly higher speed than maximum endurance and represents the speed at which the energy per unit distance is minimized or the maximum range speed.

For more elaborate assessment of flight performance, it becomes necessary to look at the lift and drag elements in closer detail, and analysis approaches that look at the distribution of forces along the rotor blade (blade element theory) are necessary.

In practical terms, a rotor blade can hurl air down at about one-sixth of the tip speed averaged over the disk. You can imagine that near the center of the rotor, the downwash speed is less; at the tip itself, it might be as much as one-third of the tip speed such that the average is one-sixth. In blade-relative terms, this means the airflow hitting the tip is diverted downward by about 20 deg, a flow rotation consistent with everyday experience with stirring tea with a spoon, or holding a hand out of a car window. Holding a wing at a shallow angle produces a lift force, which increases with angle up to some point where the flow separates behind the wing, destroying the lift and increasing drag. This lift-limiting condition is called *stall*.

So, for a given rotor diameter, we can impart a downward velocity equal to one-sixth of the tip speed, and simply increasing the tip speed (i.e., rotation rate) will increase the thrust quadratically. But we cannot go too far in this direction, because once the blade tips approach the speed of sound, shock waves can develop that increase drag and cause severe vibration. Thus, practical rotors are limited to a tip Mach number of about 0.7.

These values should be considered rules of thumb—unfortunate details may prevent them from being realized in some situations, or particularly artful design might do a little better. But until detailed examination can be made of the specifics of a given vehicle and environment, it would be imprudent to assume much higher values.

In classical aeronautics wing characteristic terms, the 20-deg turning angle corresponds to a lift coefficient of about one-third, which is particular to a specific blade shape and angle of attack at some fixed condition of Mach and Reynolds number. The lift coefficient normalizes the resulting lift force to the dynamic pressure of the flow and to the blade area, and in a way indicates how hard the wing is trying to turn the flow.

In rotorcraft engineering, it is more common to normalize forces to the disk area rather than the blade area, and so a *thrust coefficient* is used. Practical rotors typically have thrust coefficients in the range 0.01–0.05. An important bridging term is the solidity (i.e., the blade area as a fraction of the disk area). Small helicopters (like the twin-bladed Bell UH-1 Huey or the Bell 212) have rather low solidity. Large heavy-lift helicopters like the CH53-E Super Stallion with its seven-bladed rotor have a higher solidity in order to avoid blade stall. (A principal contributor to solidity is the number of blades, which is an important consideration in noise and vibration. Obviously a rotor with more blades will be heavier. The competing factors and effects account for the wide diversity of blade numbers seen in fielded aircraft.)

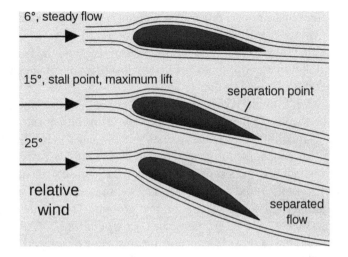

Fig. 2.7 Airflow over a wing or rotor blade. (The angles are representative, but vary for specific wing sections and flow conditions.) The downward deflection of the streamlines is what generates lift—increasing the angle of attack increases this, up to the point where the flow separates from the upper wing surface at about the quarter chord position, destroying the lift ("stall") and generating a wide turbulent wake, causing drag.
Source: U.S. Air Force.

VORTEX RING STATE AND WINDMILLING

An undesirable condition can arise in the vertical descent of a rotorcraft. Specifically, if the descent speed is comparable with the induced velocity, the aircraft is basically immersed in its own downwash and may lose lift. The flowfield will never be perfectly symmetric in any case, due to crosswind, fuselage blockage, and tip vortex effects, so flying in this condition will result in considerable buffeting and unsteadiness of lift.

The practical solution is to define the flight envelope to avoid this condition. In other words, a pilot should make a vertical descent only slowly, or descend with some adequate forward velocity such that the downwash is advected rearwards.

There are other conditions (eg, autorotation, where in a fast descent the flow provides a windmilling torque that spins the rotor up, an effect sometimes used by terrestrial helicopter pilots to land safely in the event of an engine failure), but these are not of practical use in a planetary exploration design.

GROUND INTERACTION

The rotor flowfield discussion up to this point imagines the system in free air, where the flow is unimpeded. Near the ground, however, the flow is blocked and must (for continuity of mass) be turned horizontal (see Fig. 2.8). This imposes a stress on the ground and also increases the thrust on the rotors, an important effect that must be taken into account by the pilot when taking off or landing. Many of the earliest efforts in rotorcraft were so marginal in performance that they could—just—lift off when they had the benefit of this ground effect, but in attempting to climb they lost that advantage and so were limited to heights of a rotor radius or so.

The downwash from the rotors imposes a shear stress on the ground that can cause particles to move and be lofted into the flow. If fine dust or snow is present, the particles can be lofted into a recirculating cloud where visibility becomes highly degraded, a phenomenon termed *brownout* or *whiteout*, respectively (see Fig. 2.9). Brownout became a serious problem in the intense Afghanistan desert helicopter operations in the early 2000s, leading to a number of accidents as pilots lost situational awareness near the ground. This issue stimulated considerable work on the problem, and computational fluid dynamics simulations, coupled to particle transport models, have developed to the point where reasonable predictions of the flight conditions leading to brownout may occur. The tip vortices (see next section) turn out to be a significant complication because their intense shear causes enhanced dust emission.

Dust or snow, whether naturally in the air or kicked up by downwash, can also deposit electrical charge on a flying vehicle or its rotor blades. This can also occur flying through rain, spray, or cloud; in aeronautics[¶] this is called *triboelectric charging*. Such charging is mostly a personnel hazard or a fuel

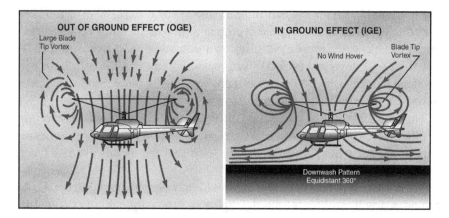

Fig. 2.8   The presence of the surface near a rotorcraft causes the downwash to spread radially outward. The flowfield modification in ground effect alters the tip vortex characteristics, and also enhances the rotor lift.
Source: Federal Aviation Administration.

¶This is somewhat confusing, because this term is used in physics to refer to the generation of charge by rubbing, rather than the deposition of charge by impinging objects, which is the sense here. In any case, there is a portrayal of the challenges posed by triboelectric charging on rotorcraft in the movie *Hunt for Red October* (Paramount Pictures, 1990).

Fig. 2.9 A U.S. UH-60 Black Hawk helicopter landing during a training exercise in Kuwait. The downwash lifts dust and spreads it radially outwards. This cloud leads to a degraded visual environment that can cause pilots to lose situational awareness and crash.
Source: U.S. Department of Defense photo by Army National Guard Spc. Jovi Prevot, May 2018.

ignition hazard (neither of which is relevant for planetary rotorcraft), but in principle, electrostatic discharges could damage electronics or cause radio interference.[3]

Tip Vortex Formation/Effects

In order to provide lift, it follows that the pressure on the upper surface of the rotor must overall be less than that on the lower surface. But at the tip of the rotor, this pressure difference causes some spillover flow from beneath the blade, around the tip toward the top (i.e., a circular or vortical flow). This results in some loss of lift (or, equivalently, introduces more drag to attain the same lift) and so is undesirable. It happens in all wings and is a reason that long, narrow wings** are generally more efficient and so are seen in sailplanes. On many modern airliners, winglets have been introduced to minimize the effect of these vortices and thus improve fuel efficiency.

The tip vortex flow is shed from the tip as it flies through the air, trailing behind and being advected somewhat downward by the downwash. The tip vortices from large aircraft can be very intense, and the persistence of them

---

**Here, *long* is meant in the sense of a large span; *narrow* means in the sense of a small chord. The ratio of the span to chord is called the *aspect ratio*, and high aspect ratio wings are efficient in having lower lift-induced (vortex) drag.

Fig. 2.10 A dramatic photo of a V-22 Osprey tilt-rotor developing powerful lift in the damp British air at the Royal International Air Tattoo in 2012. The spirals trailing from the rotor blades are the tip vortices, rendered visible by condensation caused by the adiabatic cooling in the low pressure cores of the vortices.
Source: Peter Groneman. BY-SA-2.0.

(they take some minutes to dissipate) is a factor in determining how close aircraft landings or takeoffs can be safely arranged by air traffic control.

For rotorcraft, the tip vortices can be a significant consideration in a number of domains. The intense shear in the tip vortex flow can make it effective in picking up particulate material from the ground, causing sandblasting effects or generating a cloud of dust (much more so than if the rotor downwash were merely a column of downflowing air as in the actuator disk idealization).

Tip vortices are also significant for vibration and noise. Strong increase in the clattering noise of twin-rotor Chinook helicopters occurs in flight conditions when the rear rotor chops through the vortex system of the forward rotor. Similar effects can occur in multirotor vehicles and lead to aerodynamic performance losses and higher vibration levels.

Occasionally, tip vortices are rendered visible by condensation (see Fig. 2.10). This occurs because the upper surface of the rotor, as well as the vortex tube that trails from the tip, are at low pressure. Ambient air, having suddenly been expanded into this low pressure region, is adiabatically cooled. If the air is moist enough, this thermodynamic temperature drop may reach the dew point, and water vapor condenses into small fog droplets. The onset of such condensation fogging can also be seen on fixed-wing aircraft, such as from the overwing window seat of an airliner on moist days at takeoff or landing. These are moments when the suction and thus lift coefficient are particularly high; therefore, the wing is working harder.

FORWARD FLIGHT: ADVANCE RATIO

The flow around a vehicle hovering in still air is, broadly speaking, symmetric (tail rotor and fuselage blockage effects notwithstanding), and the lift on the rotor is similarly symmetric (but not uniform—lift tends to be concentrated toward the faster-spinning blade tips). This symmetry breaks down in forward flight. If the vehicle is flying forwards at a speed V, and the blades are spinning with a tip velocity V_T, then the advancing blade will encounter the air at a velocity (V_T+V), whereas the receding blade hits the air more slowly, at $(V_T - V)$. The ratio (V/V_T) is termed the *advance ratio* (see Fig. 2.11).

The effect is a fundamental limit on helicopter flight speed, because lift depends on the local airspeed squared; thus, all else being equal, the lift on the advancing side of the rotor will be larger than that on the receding side, generating a rolling torque that must be corrected.

At the vehicle level, advance ratio effects can balance out on a multirotor vehicle; however, the varying forces over the course of each blade rotation will

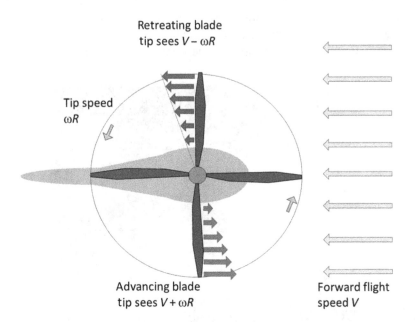

Fig. 2.11 Advance ratio. Forward flight is ultimately limited by the superposition of the forward speed with the rotor tip velocity ($V_T = \omega R$), which makes the airspeed seen by the blades very asymmetrical. Cyclic control can only counter this effect so much; eventually the trailing blade cannot develop enough lift (stalls), whereas the leading side may encounter Mach number (compressibility) effects.
Source: Author.

induce undesirable vibration and fatigue loading. In traditional helicopters, these effects are mitigated by manipulating the rotor blade geometry over the course of a rotation. Planetary applications have not been too concerned with advance ratio effects, in that forward flight speeds that have been attempted or contemplated so far have been modest compared with the tip speed.

CONTROL: HELICOPTERS

The lift generated by a blade depends on its angle of attack. By pitching up, more lift is generated. Thus, if the pitch of all the blades of a rotor is increased, the rotor thrust goes up. This action is performed in helicopters via a set of linkages pushed via a swashplate (see Fig. 2.12). The modification of pitch of all the blades on a rotor together is called *collective control*; on a piloted helicopter this movement is effected by a lever that is moved up and down. The collective lever usually also has the engine throttle control mounted on it, because a change in collective pitch changes the rotor torque, and the throttle setting must therefore also be changed to balance this while maintaining a constant rotor speed.

To make translational (forward or sideways) flight, the thrust axis of the rotor must be tilted (see Fig. 2.13). This is performed via the same swashplate, but by tilting it rather than sliding it up and down for the collective control. The swashplate tilt, via linkages to the blades, causes the pitch of each blade to vary cyclically over the course of one rotation; this, in turn makes the lift

Fig. 2.12 A simple swashplate assembly. As the rotor spins, the red pitch linkages that slide across the tilted plate cause the blades to change their angle of attack and thus cyclically change the amount of lift. If the entire plate is lifted up or down, the overall (collective) amount of lift changes.
Source: Computer rendering by KoenB, public domain via Wikimedia.

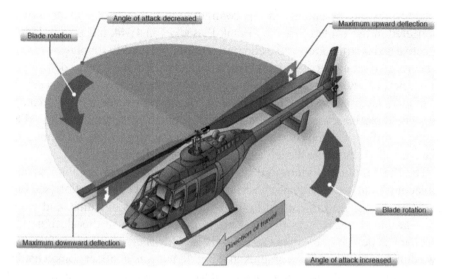

Fig. 2.13 Forward and lateral control in a helicopter is effected by tilting the rotor plane with the cyclic control. That tilt involves changes in the blade angle of attack throughout the rotation. These angle changes, coupled with the gyroscopic effect of the rotor angular momentum and the relative airspeed changes over each rotation due to forward flight, necessitate considerable complications in the rotor hub, which must accommodate and control flapping and lagging (forward/back) motion of the blades.
Source: Federal Aviation Administration.

distribution asymmetric. The tilt of the swashplate is commanded by a cyclic control stick.

The spinning rotor blades experience drag, which is countered by the engine torque; however, that engine torque, in turn, reacts against the vehicle, which would tend to spin in the opposite direction to the blades, so the drag torque must be balanced somehow. A number of helicopter configurations have been successful with different approaches to solving the torque balance problem. The conventional single-rotor design uses a horizontal tail rotor (or occasionally[††] a ducted fan called a *fenestron*) to balance the drag torque on the main rotor (see Fig. 2.14). The power that is expended in driving the tail rotor is in effect wasted, and indeed the tail rotor is often also a major source of noise. Some designs, especially those for high speed, have large tail fins such that in forward flight the fin can provide some stabilization. Because the thrust from the tail rotor can be adjusted quickly over a wide range (by changing its collective pitch via control pedals), this configuration is rather agile in yaw.

The second major configuration type is the twin rotor. Side-by-side configurations like the Fw-61 have been successful, and in fact the largest rotorcraft flown, the Mil V-12, was also of this type. Tilt-rotor aircraft, like the

---

[††]An additional, but rather rare, variant uses air blown through a slot in a cylindrical tail boom to effect yaw control via the Coanda effect.

V-22 Osprey, are essentially in this configuration in hover. The tandem twin, pioneered in the United States by Frank Piasecki[‡‡] in 1944, has proven highly effective and is structurally more compact. Despite the need to supply power with synchronized rotation to rotor hubs at the front and back of the vehicle, demanding a complex, heavy engine transmission, the configuration enjoys much success in its modern incarnation, the CH-47 Chinook. Because the tandem rotors rotate in opposite directions, their drag torques balance out. Yaw control is effected by applying different cyclic control to the two rotors (eg, to yaw left, the front rotor plane is rolled slightly to the left and the rear rotor to the right).

The third main configuration[§§] is the coaxial rotor, as in Lomonosov and Amecourt's early experiments. This is a rather compact configuration (see Fig. 2.15), well-suited to operation in the cramped quarters on ships, and many Russian naval helicopters by the Kamov corporation use this design. The mechanical design of the concentric shafts, and the accommodation of the two swashplates, is somewhat complex and bulky, but this is offset somewhat by the elimination of the tail rotor. Yawing of a coaxial helicopter is accomplished by differential collective control, shifting the balance of lift and drag from one

Fig. 2.14   A French army Aérospatiale SA 341F2 Gazelle helicopter in the desert during operation Desert Shield. The helicopter has the traditional single-rotor layout, but the tail rotor is of the embedded fenestron type.
Source: U.S. Defense Imagery VIRIN: DF-ST-92-07419 by TSgt. H.H. Deffner.

[‡‡]His company was subsequently renamed Vertol, and then was later acquired by Boeing.

[§§]Arguably there is a fourth distinct configuration, although it could be considered a special kind of twin. This is the synchropter, which has two slightly canted top-mounted rotors. The rotor disks cross, but by phasing the (usually two-bladed) rotors with appropriate gearing, they can mesh together. This provides a rather compact arrangement, although one I find aesthetically unappealing. Used on the first production helicopter, the Flettner Fl 282 Hummingbird, it has found only very rare use since (eg, on the Kaman Huskie and K-Max). Anton Flettner was brought to the United States from Germany in Operation Paperclip and became the chief designer at the Kaman Corporation.

Fig. 2.15 A modern coaxial-rotor helicopter, the Kamov Ka-27 Helix flying from a Russian destroyer ship. Note the wheeled undercarriage and chin-mounted radome. The fins at rear provide stability during forward flight.
Source: U.S. Department of Defense Photo by Jason R. Zalasky, U.S. Navy.

rotor to the other, such that its drag torque becomes dominant. The Mars helicopter is of this configuration.

Some helicopters have avoided the complexity of flapping and lagging hinges by accommodating the necessary motions by elastic flexure, rather than sliding. The successful Bo-105 is one example.

CONTROLLABILITY: MULTIROTORS

Control of a multirotor vehicle by rotor speed control alone, without changing the geometry of the rotor blades, is mechanically and in some ways conceptually simpler than cyclic control.[¶] However, it begs the question (which arose early in the formulation of the Dragonfly mission): Why we do not see more multirotor vehicles of this type carrying people in the last century? The reason, at least in part, is because of response time.

To respond to a gust, for example, a vehicle must increase the thrust on one side by some amount to pivot the vehicle such that the thrust can balance not only the weight of the vehicle in hover, but also the side force due to the gust. Because the thrust depends on the rotation rate of the rotor, it follows that the rotor speed increase must be rapid.

The moment of inertia of a rotor of fixed mass goes as the square of its diameter or, for a constant shape and density, as the fifth power of the diameter.

[¶]There is an intermediate multirotor architecture, seen for example in the De Bothezat vehicle, where there is a collective (but not cyclic) control for each of the rotors, which spin at a constant speed. Even this, of course, was a challenge for the pilot to manage.

Thus, as vehicles of progressively larger scale are considered, the rotor inertia and thus the response time to commanded changes in rotation rate grow very rapidly. While small hobby drones (Fig. 2.16) can twitchily respond with rotors that change speed in a fraction of a second, if you observe a typical helicopter starting up, you will note that it takes tens of seconds for its large main rotor to reach its operating speed. Angular acceleration is the ratio of torque to moment of inertia, so a faster response can be attained by having a more powerful motor, but this undesirably increases its mass. The rotor and hub structures must also be strong enough to take the larger forces and torques, and the speed control electronics and wiring for the motor must also be more capable (and heavy). Thus, there are practical limits on the rotor diameter (maybe 1.5–2 m) for this type of control.

Fig. 2.16 A popular camera quadcopter drone. It is obvious that increasing the thrust on the two forward rotors will cause the nose to rise—in some respects the control of a multirotor vehicle is more intuitive than that of a helicopter. This model (DJI Mavic) has rotors and rotor arms that fold into a compact configuration for transport, and its stabilized camera view is broadcast to a cellphone screen mounted in a headset. Here the drone is being used to carry a sensor package (blue box) into dust devils on a desert playa to measure their dust content and vortex intensity.[4]
Source: Author.

Fig. 2.17 Ehang 216 two-person multicopter. The compromise between efficient lift performance (preferring large rotors) and control response and rotor footprint issues (preferring smaller rotors) yields a vehicle with eight pairs of ~1.5-m rotors.
Source: Ehang news release.

The advent of person-carrying electric multirotor vehicles (see Fig. 2.17) has become possible in only the last half-decade or so by the combination of lightweight, efficient, and powerful rotors, motors, drive electronics, and batteries. The emergence of electric cars has doubtless been a significant factor in driving some of these developments. Even so, the strong dependence of rotor inertia on diameter represents a fundamental challenge to large multirotor vehicles, with the result that vehicles with enough thrust to carry more than one person typically have many more than four rotors (such that their individual diameter, and thus response time, can be small), or they use variable blade pitch at more or less constant rpm to effect control.

Conceptually, the use of differential thrust on a quad-rotor vehicle to perform roll and pitch is intuitive—the appropriate rotor or pair of rotors is throttled up to raise the desired side of the vehicle. In the case of an X configuration, pairs would be used for the relevant cardinal direction. In the case of a + configuration, a single rotor change might be used. In either case, the response is not a pure pitch or roll, because changing the speed of one rotor changes the net drag torque on the rotor blades, such that a slight yaw acceleration would be introduced.

This cross-coupling is the source of difficulty of learning to fly rotorcraft, and doubtless is the cause of many pilot failures, where a control input causes effects in multiple dimensions in a way that is much less intuitive than for a fixed-wing aircraft. The gyroscopic stiffness of a large spinning rotor introduces other unintuitive cross-coupling effects. However, for a computer-controlled autopilot, the coupling of aerodynamic and gyrodynamic effects is

simply the choice of the correct values in the *plant matrix*, an array of numbers
that maps control inputs to dynamic results. In an ideal system, this matrix is
mostly zeros, and a single diagonal line of nonzero numbers is the scale factor
that maps the change in a rotor rpm (say) or the setting of a rudder to the
vertical acceleration or yaw rate. In a more complicated system, many of these
cross- or off-diagonal terms are nonzero, such that the rotor rpm change
causes a yaw as well as a vertical climb.

A pilot adapting to a new aircraft is, in essence, attempting to become
familiar enough with the terms in this matrix that the response of the vehicle
can be predicted and thus the control inputs required for a desired maneuver
are learned. The trick, of course, is to learn these terms (an instinctive version
of what engineers call *system identification*) without encountering some
physical divergence condition in the process that results in a crash!***

Because a computer autopilot can be programmed with (or even programmed
to learn) the appropriate plant matrix, it becomes straightforward to indulge
propulsion and control designs that have large cross-terms. Thus, although the
human pilots of the de Bothezat quadcopter wallow around, compensating for
gusts and the unexpected cross-coupling of their controls, a computer can
deftly and instantly adjust the speed settings of its different rotors.

Thus the emergence of multirotor hobby drones, while also enabled by
suitably lightweight lithium batteries, was largely due to the availability of
small sensors and the computer control to stabilize flight. If a human pilot had
to do differential throttling of the four engines on a small quadrotor directly
by radio control, the process would be hopeless (worse, in fact, than for a
full-scale quad, in that the dynamical evolution would be much more rapid
due to the small vehicle inertia). However, the human operator merely
provides a command that is translated by the fly-by-wire autopilot into the
relevant speed control changes. And the evolution from that architecture,
wherein the command of where to fly or how fast is not provided by a human
in real time but is instead one of a preprogrammed sequence of flight segments
or is generated automatically from some sensor to track a target or hold a fixed
position relative to it, is straightforward. Hence (with parallel improvements
in motor and battery technology), the drone revolution.

Note that although yaw control can be effected on quadcopters with
perfectly plane (horizontal) rotors by differential rotation, the yaw torques
available (due to rotor drag) are rather modest on full-scale vehicles,††† and so
yaw authority is weak. This can be mitigated by canting the rotors inward so
their thrust has a component in the horizontal plane. Again, this cross-coupling

---

***A useful tool in this process is the flight simulator, which allows the pilot to become familiar with
the response to control inputs without exposure to the hazard of a departure from controlled (real) flight.

†††Actually, at the low Reynolds numbers typical for small drones, the profile drag coefficient of the
rotors is comparatively high, and so the yaw authority of horizontal planar rotors is often adequate for
hobby-type drones.

of control axes would be initially daunting for a human pilot, but for an autopilot this simply requires an adjustment to the cross-terms in the matrix.

NOISE AND VIBRATION

Although not central to the challenge of getting off the ground initially, considerations of noise and vibration are important in rotorcraft design, because the large rotating elements generate fluctuating structural loads and pressures. Having done a couple of TV shoots being interviewed in helicopters, I can attest that they are noisy![‡‡‡]

Although passenger comfort is not a factor in planetary rotorcraft, questions of structural stiffness and fatigue are. Furthermore, scientific instrumentation may be susceptible to vibration, corrupting data. As an example, on Titan where solar illumination is rather weak, the exposure times for camera images must be longer than on Earth; if structural vibration is excessive, images could become motion-blurred.

The most important parameter is the blade passage frequency (i.e., how many times per second a blade passes a given point or direction). A two-bladed rotor at 600 rpm will have a blade passage frequency of 20 Hz, and this frequency (and the harmonics of it) will be that at which the vehicle structure will be most strongly excited.[§§§] If elements of the structure have natural modes near these frequencies, then those modes will be resonantly amplified. Thus, helicopter design often involves tuning to avoid modes at these frequencies. It is, of course, conceivable to make the structure exceptionally stiff, raising the natural modes to frequencies well above those of the rotor, but this is usually expensive and heavy, and so the helicopter designer must make do with just enough stiffness to nudge principal modes out of the resonant frequencies. For conventional helicopters (with a relatively constant nominal rpm), this is a better-posed problem than for multirotor vehicles with more widely variable speeds. On the other hand, the structural loads imparted by an individual (smaller) rotor on such vehicles may typically be less than on a helicopter. As ever, elegant design becomes a question of balance between competing effects.

A particularly destructive interaction is called ground resonance, which can occur when the rotors cause amplification of oscillation in the landing gear. During rotor spin-up or spin-down on the ground, lower frequencies than normal will be encountered. Also, wind can cause appreciable displacements of the rotors (blade sailing) as if cyclic input had been made. This is particularly a problem for less stiff rotor blades (typically larger ones), which rely on the centrifugal force of rotation to keep them taut and straight (see Fig. 2.18).

[‡‡‡]Much of the noise, of course, is merely due to the presence of a very powerful gas turbine engine just a few feet from the cabin, rather than the structural or aerodynamic aspects of rotorcraft specifically.

[§§§]A 1/rev frequency will be excited if the rotor is unbalanced somehow, such as by damage to a blade.

Fig. 2.18   The largest helicopter ever built, the Mil V-12 Homer, created to carry ballistic missiles to their remote silos, could carry 196 passengers or some 40 t of cargo. Its two rotors are 35 m in diameter, and the rotor blades have perceptible droop.
Source: Author, at the Monino Air Force Museum near Moscow.

Such rotors are often seen to droop when still, and the rotor disk may bow upward when high thrust is generated, such as at takeoff.

GUIDANCE AND NAVIGATION

Multirotor speed adjustments or the adjustment of a swashplate are only a means to an end, namely getting the vehicle to where you want it to go. That which you do not measure, you cannot control. Thus, an essential part of the drone revolution, and of planetary rotorcraft, is the ability to sense position, orientation, and velocity relative to the ground.

This sensing was implemented in early helicopters simply and instinctively by the pilot's eyes, and human vision remains the principal tool in clear daytime conditions. However, in certain applications, such as for all-weather or night low-level flying demanded by some military operations, or in conditions of whiteout or brownout, sensing systems to supplement or replace the pilot's own senses have been introduced in aircraft. These capabilities have been necessary for lunar or planetary landing from the very beginning of the space age, however.

The first soft-lander to the moon, the NASA Surveyor I in 1965, used a radar altimeter to trigger the firing of its large solid rocket motor to remove the kilometers-per-second arrival velocity. (This function is conveniently performed by the atmosphere on Mars or Titan.) Then a multibeam Doppler radar (see Fig. 2.19) was used to sense the vertical and horizontal velocity relative to the lunar surface. Similar sensors were used by Apollo.

Radars can give occasionally spurious data (eg, one Surveyor radar triggered briefly on the spent casing of the solid rocket as it fell away, and the Phoenix Mars lander radar pinged its heatshield), so it is important to filter the data with a navigation process. A typical input into such a process is an inertial measurement unit (IMU), a combination of accelerometers that measure change of speed and angular rate sensors that measure rotation.[¶¶¶] An IMU basically provides a means of keeping track of position, velocity, and orientation changes from a starting point, but because it computationally

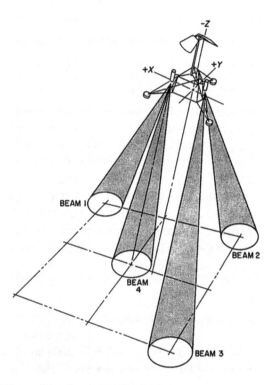

Fig. 2.19 A multibeam Doppler radar, like the Surveyor system shown here, measures the line-of-sight velocity in three or four directions, which can be processed to yield the ground-relative horizontal velocity components and the vertical velocity.
Source: NASA/Jet Propulsion Laboratory (JPL).

¶¶¶Originally, IMUs were developed for ballistic missile guidance, and the angle sensors were arranged to maintain a platform in an inertially fixed orientation. The output of the three orthogonal accelerometers on this platform could then be integrated twice to derive the position change in that inertial frame of reference. Although computationally simple, this platform and the mechanical gyroscopes to sense its orientation demanded exquisite precision engineering: the Apollo lunar module used an IMU of this type. This architecture has been progressively supplanted by a mechanically simpler alternative, the strapdown IMU, in which the angle sensors (now typically solid-state resonators or ring laser gyros) and accelerometers are just rigidly fixed to the vehicle, the angle sensor output is integrated to keep track of the vehicle orientation to the inertial frame, and the accelerometer outputs are mathematically rotated into that frame. The additional computational complexity is now relatively trivial compared to the mechanical simplification and lifetime/reliability improvement, and so this architecture is now almost universally used. The fundamental limitation, that navigation requires regular external updates to correct for IMU drift, remains.

integrates position and speed from that point, uncertainties grow progressively with time. Thus, in most applications, an IMU will drift unacceptably over the mission, and the navigation estimate must be updated by some external reference such as a compass or star tracker (for orientation), or radar, optical (or, for submarines, sonar) velocity, or position fixes. An IMU, however, is a powerful contributor to maintaining an accurate estimate between such fixes. Also, although the IMU position estimate degrades quickly, the orientation estimate of an aerospace-grade IMU[****] is typically very good indeed, and for a mission of a few minutes or tens of minutes, good enough not to require external update. Thus, modern Mars landers, and drones, do not require optical horizon sensors or other attitude references to fly upright; an inertial reference at the takeoff point is good enough. (Note that the artificial horizon in classical aircraft cockpits is effectively an inertial reference, provided by a gyroscope.)

A radar altimeter, an IMU, and a multibeam Doppler radar were used on the Viking landers. Although Mars Pathfinder in 1997 triggered a braking rocket with radar, it accepted the risk (a modest one given its small mass and thick airbags) that winds might cause a high horizontal velocity at landing. The more massive Mars Exploration Rovers (MERs) in 2003, however, found that the tolerance of their airbags for horizontal motion at impact was less, and a rocket system to null that motion had to be installed. A Doppler radar, however, was a heavy and expensive item to add, especially at a late stage in design, but by now imaging technology had advanced to the point that a short sequence of optical images could be processed in real time to solve for the translational and rotational motion of the imaging platform. Thus the MERs used the Descent Image Motion Estimation System (DIMES)[5] to derive the ground-relative velocity.

For true soft landing on Mars, however, a Doppler radar was still needed for the Mars Science Laboratory (MSL, aka Curiosity), whose design started in the mid-2000s; the Phoenix and InSight landers of 2008 and 2018, respectively, also used radars. But lunar lander developments in the United States in the 2010s, such as Mighty Eagle,[6] had begun to assume optical navigation techniques.

The Japanese Hayabusa spacecraft obtained an asteroid sample via a touch-and-go landing in 2005, using optical sensing to determine its horizontal velocity. It did so, however, by deploying a highly reflective marker onto the surface to serve as a reference, making the feature extraction image computation rather easier. The marker was a sort of beanbag, to ensure it would not bounce on the asteroid surface.[7] The Chinese lunar lander Chang-E-3 in 2015 used an optical imaging system.[8] Now many consumer drones use optical sensing, too.

[****]A typical aerospace IMU for missile or spacecraft use weighs a couple of kilograms, may cost about $1 million, and is a strategic item governed by export control laws. Cellphones have small IMUs about the size of a penny, and a cost not much more than that, but their errors grow far more quickly.

At the beginning of this section, we noted the need to sense the vehicle velocity relative to the ground. In fixed-wing aviation, velocity relative to the air is the paramount consideration, because lift relies on this. For most rotorcraft, however, wind is merely an error term in the navigation estimate. Measuring wind speed on board separately is not easy, because the flowfield around the vehicle is obviously strongly perturbed by the rotors, but in principle could help with gust compensation to achieve tighter station-keeping relative to the ground.

Although the earliest rotorcraft pilots relied only on their eyes for position, speed, and orientation control, this is inadequate in conditions of poor visibility or at night, or over tracts of relatively featureless forest or desert. Helicopters for the deployment of special forces at night, for example, are often equipped with radar systems for precision ground-relative altitude and speed (and for hazard detection). IMUs have been used for long-range navigation, although since the 1980s have been somewhat supplanted by Global Positioning System (GPS) satellite signals.

Thus, both planetary landers and helicopters have seen a range of solutions to the guidance and navigation problem. Suffice it to say there must be a suite of sensors and a means of deriving the vehicle state (position, attitude, and velocity) from them, and to act accordingly.

This chapter has given an overview of the problems of atmospheric flight that is cursory in the extreme. If it has only succeeded in indicating how complicated a challenge it can be, then it has been successful.

REFERENCES

1   Popular helicopter texts in the United States are the inexpensive Dover editions of Wayne Johnson's *Helicopter Theory* (1980) and Stiepniewski and Keys's *Rotary Wing Aerodynamics* (1984). Although excellent for their intended purpose and audience, I personally find these somewhat parochially focused on the problems of conventional helicopter design, and the use of Imperial units throughout (as traditional in U.S. aerospace engineering) is a distracting self-imposed handicap. Simon Newman's *The Foundations of Helicopter Flight* has good exposition with simple models of the mechanical aspects of helicopters, and an interesting appendix on windmills. I find Gordon Lieshmann's text, *Principles of Helicopter Aerodynamics* (Cambridge, 2006) to be by far the most useful book for the things I have wanted to know about: its treatment of tip vortices is excellent. A nicely accessible treatment of helicopter flight from a pilot's perspective is W. J. Wagtendonk's *Principles of Helicopter Flight* (ASA, 2019), which covers the basics of rotor mechanics (essentially without equations) but also such topics as underslung loads, icing, mountain flying, and the like. Much of the material covered in that volume is also treated in the publication *Helicopter Flying Handbook* (FAA-H-8083-21B) that is free to download from the Federal Aviation Administration at https://www.faa.gov/regulations_policies/handbooks_manuals/aviation/
2   Lorenz, R. D., Claudin, P., Radebaugh, J., Tokano, T., and Andreotti, B., "A 3 km Boundary Layer on Titan Indicated by Dune Spacing and Huygens Data," *Icarus*, Vol. 205, 2010, pp. 719–721. See also Lorenz, R. D., and Zimbelman, J., *Dune Worlds: How Wind-Blown Sand Shapes Planetary Landscapes*, Praxis-Springer, 308 pp, 2014.

3   See Rabinovitch, J., Lorenz, R., Slimko, E., and Wang, C., "Scaling Helicopter Brownout for Titan and Mars," *Aeolian Research*, Vol. 48, 2021, 100653, and Lorenz, R. D., "Triboelectric Charging and Brownout Hazard Evaluation for a Planetary Rotorcraft," AIAA-2020-2837, *AIAA Aviation Forum 2020* (virtual). The triboelectric charging effect was also considered in Farrell, W. M., McLain, J. L., Marshall, J. R., and Wang, A., "Will the Mars Helicopter Induce Local Martian Atmospheric Breakdown?," *The Planetary Science Journal*, Vol. 2, No. 2, 2021, p. 46.

4   Jackson, B., Lorenz, R., Davis, K., and Lipple, B., "Using an Instrumented Drone to Probe Dust Devils on Oregon's Alvord Desert," *Remote Sensing*, Vol. 10, No. 1, 2018, p. 65.

5   Cheng, Y., Johnson, A., and Matthies, L., "MER-DIMES: A Planetary Landing Application of Computer Vision," *2005 IEEE Computer Society Conference on Computer Vision and Pattern Recognition (CVPR'05)*, Vol. 1, 2005, pp. 806–813, doi: 10.1109/CVPR.2005.222. Note that because of hardware limitations on the cameras, the image sequence spanned several seconds, so IMU data were needed to help determine the velocity solution. A foundational paper on the overall problem is Johnson, A., Ansar, A., Matthies, L., Trawny, N., Mourikis, A., and Roumeliotis, S., "A General Approach to Terrain Relative Navigation for Planetary Landing," *AIAA Infotech@ Aerospace 2007 Conference and Exhibit*, p. 2854.

6   McGee, T. G., Artis, D. A., Cole, T. J., Eng, D. A., Reed, C. L., Hannan, M. R., Chavers, D. G., Kennedy, L. D., Moore, J. M., and Stemple, C. D., "Mighty Eagle: The Development and Flight Testing of an Autonomous Robotic Lander Test Bed," *Johns Hopkins APL Technical Digest*, Vol. 32, No. 3, 2013, pp. 619–635, https://www.jhuapl.edu/Content/techdigest/pdf/V32-N03/32-03-McGee.pdf

7   Hashimoto, T., Kubota, T., Kawaguchi, J. I., Uo, M., Shirakawa, K., Kominato, T., and Morita, H., "Vision-Based Guidance, Navigation, and Control of Hayabusa Spacecraft—Lessons Learned from Real Operation," *IFAC Proceedings Volumes*, Vol. 43, No. 15, 2010, pp. 259–264.

8   Li, S., Jiang, X., and Tao, T., "Guidance Summary and Assessment of the Chang'e-3 Powered Descent and Landing," *Journal of Spacecraft and Rockets*, Vol. 53, No. 2, 2016, pp. 258–277.

Chapter 3

Challenges of Space

"Anyone who sits on top of the largest hydrogen-oxygen fueled system in the world, knowing they're going to light the bottom, and doesn't get a little worried, does not fully understand the situation."

—John Young, American astronaut

Rotorcraft engineering and spacecraft engineering are thoroughly distinct disciplines. Although modern aircraft are replete with electronics and some of the same systems engineering approaches are needed to manage complex projects, the main challenges in helicopters typically relate to the mechanical considerations of structures and dynamics—to deliver torque to a rotor and to control the vehicle. Vibration and aeroelastic effects are of paramount importance. But the presence (historically) of a human pilot, and the opportunity to repair or redesign in response to observed performance—perhaps from thousands of flights in hundreds of airframes—introduces feedback into the engineering process.

Spacecraft engineering has arguably a more electrical emphasis. Although the mechanical aspects of spacecraft are by no means undemanding, the defining feature of systems in space is their remote operation, and so communications and automatic control were key disciplines in the dawn of the space age. A space system is often a one-off custom design; once built and launched, it is never seen again and must work the first time. The ability to predict how the system will perform in what are very unfamiliar environments is crucial, and hard-won lessons[1] from the early days of space exploration demand exceptional rigor in the construction and testing of space hardware.

This book is not the place for a detailed discussion of spacecraft design,[2] but it is appropriate to summarize some aspects of the ground, launch, and space environments that a planetary rotorcraft must be designed for.

THERMAL VACUUM

Space by definition is (virtually) empty. As such, space doesn't have a temperature in that the thermodynamic temperature of a gas is a statistical expression of the velocities of the molecules, and there are too few interactions of molecules for this to have much practical meaning.

The temperature of objects in space, then, is governed by the flow of heat to and from them, determined principally by radiation. Thus the surface of an insulator placed near the sun so that it fills the sky will nearly attain the temperature of the sun's photosphere (~6000 K), whereas if it is shaded from the sun and sees only deep space, it will cool towards the cosmic background temperature of about 3 K. Most surfaces, including that of our planet, are somewhere in between these extremes. Spacecraft engineers must use coatings like metal foils or special paint to control the exchange of heat by radiation to maintain benign conditions. In some cases, spacecraft are slowly rotated to even out solar heating ("barbeque roll"). In other cases, a spacecraft may have a "hot side" and a "cold side", and measures such as conductive heat paths or even heat pipes or pumped fluid loops are used to spread heat around. Special attention must be paid to situations like eclipses where a transient cold condition may occur.

Any components (e.g. power-switching transistors or radio amplifiers) where significant power dissipation occurs may be prone to overheating, since the convective cooling that we take for granted in the Earth's atmosphere does not occur. Equipment (first at the component level, but ultimately the whole spacecraft – e.g. Fig. 3.1) must be tested in vacuum conditions to ensure that hot spots do not develop and threaten the reliability of electronic or other parts.

MATERIALS FOR SPACE

The vacuum of space allows volatile materials to migrate, restricting the use of some polymers, adhesives, and lubricants. A standard protocol in approving materials is that under specified conditions of vacuum and temperature (125°C/257°F), they must lose less than 1% of their mass over a 24-h period; in addition, less than 0.1% of that mass can consist of condensable volatile materials (typically organics like plasticizers or solvents; water is not included in this allowance). Nylon is a common polymer that violates these rules, so it is not used structurally. (A waiver is often invoked for parachutes for which nylon is often the best material, and special bakeout procedures may be invoked to drive off moisture and other volatiles.) A general concern with molecular contamination (such as skin oil in fingerprints, or condensed volatiles) is that exposure to the strong ultraviolet radiation in space can darken the materials, interfering with the performance of optical sensors or solar cells, or with the performance of white thermal control paints. In scientific situations, especially relating to the possibilities of prebiotic

Fig. 3.1 Thermal/vacuum testing is a central part of spacecraft qualification. Here the massive Cassini spacecraft (with Huygens probe in gold-colored blankets) is in the 85-ft (26 m) tall Space Simulation chamber at the Jet Propulsion Laboratory (JPL). In this test, powerful lamps at the top are being shone on the white high-gain antenna (which Cassini used as a sunshade in space). The dark walls of the chamber may be chilled with liquid nitrogen to simulate the cold of deep space. Note technician at lower left for scale. Source: NASA.

chemistry or traces of past life, organic contaminants may give undesirably high background levels of compounds of interest. Components are usually handled with gloves and cleaned with solvents to remove any organic traces.

Particulate contamination is also important to avoid, so spacecraft hardware is assembled in controlled-access clean rooms with filtered air conditioning. Booties and sticky mats limit the ingress of dirt on shoes, and smocks or bunny suits with hats are worn to avoid dust from clothing, skin flakes, and the like (Fig. 3.2).

Tribology—the discipline of measuring and controlling friction—is a challenging area. Typically, fluorocarbon oils or greases with very low vapor pressures are used as lubricants, although attention must be paid to the operating temperature range. Occasionally solid lubricants such as molybdenum disulfide are used (requiring special deposition techniques), although the lifetime of such coatings can be limited. Although graphite is an excellent solid lubricant in the terrestrial atmosphere, its effective sliding relies on

Fig. 3.2 Mars helicopter on the belly pan mounting plate that will attach it to the Perseverance rover. The bunny suit with gloves and head coverings minimizes the deposition of skin oils, dandruff, and other contamination. Note that cuffs seal the wrists. Blocks of yellow foam provide some slightly compliant support to the rotor blades to protect them from the vibration of launch. The curved protective cover for the helicopter is just visible behind the technician's shoulder, inside a thermal test chamber whose green door is open. Source: Lockheed Martin.

adsorbed water vapor, which is released in vacuum. Graphite is thus an ineffective lubricant in space. This subtle but profound change is a good example of the surprising effects that occur in the space environment. Note that in space vacuum, bare surfaces of the same metal can cold weld together, so unlubricated contacting surfaces must be considered with care.

Certain metals, such as tin or cadmium, are also not permitted (or at least require carefully documented and justified exceptions). These are often used as a corrosion-resistant plating in terrestrial hardware, but the materials can migrate in vacuum, forming tiny whiskers over time. These whiskers can short out electrical circuits. Similarly, brittle intermetallic compounds ("purple plague") can form between gold and aluminum, so care in the selection of wiring is essential. (An entire fleet of Soviet missions to Mars was lost in 1972 because a certain model of transistor had wire bonds that allowed such junctions to form and break.)

RADIATION

On Earth, we are protected from energetic particle radiation from the sun and from deep space by both our thick atmosphere and the Earth's magnetic field.

Radiation effects on space hardware manifest in three principal ways.* First are total dose effects: Energetic radiation lancing through materials can displace atoms or cause local chemical changes. In extreme cases, materials properties such as strength or plasticity can be degraded. Similarly, darkening can occur in some transparent materials such as optical fibers or windows, degrading their function. For some electronic parts, total dose effects may cause parameter changes such as a shift in gate voltage or an increase in leakage currents.

A second type of effect is transient, called a single event upset (SEU). Effectively, it is a random "blip" caused by the charge deposition by a particle, and it is mostly a concern for computer memories, especially those storing program instructions. An SEU can change a binary 0 to a 1 or vice versa. In many cases this may be inconsequential (a spurious white pixel in an image, for example), but in just the wrong place in memory this could cause a program to crash or even trigger some unwanted event like a thruster firing (see Fig. 3.3). The usual protection is to introduce error-correction coding into computer memories (redundant bits) and have a background routine to repair errors faster than they can accumulate.

The third effect is called latchup (or sometimes single-event latchup, SEL), where a cosmic ray produces a conductive path (in effect, a parasitic thyristor) in the semiconductor matrix. This conductive path acts like a short circuit and

Fig. 3.3 Ultraviolet images of the sun taken by a camera on the SOHO spacecraft during a massive solar flare in 2003 illustrate space radiation. Energetic protons launched from the sun reached the spacecraft 12 h later, causing many spots and lines where the particles changed the charge on the detector pixels. If these were computer memory cells rather than image pixels, the instructions could be corrupted, causing failures.
Source: SOHO Extreme UV Imaging Telescope, courtesy European Space Agency and NASA.

---

*Actually, there is a fourth, *deep dielectric charging*, in which high fluxes of charged particles can deposit large amounts of charge inside materials, eventually causing electrical breakdown or arcing. This effect is encountered in only exceptional circumstances (e.g., high-altitude Earth satellites during solar storms or in Jupiter's intense radiation belts).

can impede the operation of the device, or in exceptional cases may allow sufficient current to flow to cause local heating and burn the device out. There are some electronic fabrication approaches (such as silicon-on-sapphire) that are resistant to this effect, and special-purpose microprocessors for space applications use these methods. In other cases, careful circuit design can limit current flow to avoid permanent damage, or if a current increase is detected, briefly disconnect the device to clear the SEL. (Yes, turning it off and on again is the fix for space radiation!)

The vulnerability of individual components to these different effects is not always easy to predict, and even minor changes in design or fabrication process can change the radiation susceptibility. Thus rigorous testing is necessary, and space agencies maintain lists of approved parts whose reliability has been demonstrated. The range of parts available to a spacecraft designer is far smaller than that available for terrestrial application.

LAUNCH

Getting from the Earth's surface into interplanetary space entails accelerating it to speeds above 11 km/s (25,000 mph). That process, executed by rocket launch vehicle (Fig. 3.4), must be rather rapid in order to be efficient

Fig. 3.4 A Titan IV launcher lifting off in 1997 to hurl the 5-t Cassini-Huygens mission toward Saturn. It is impossible to convey on the static printed page the deafening fury of a rocket vehicle that stresses its payloads with acceleration, vibration, noise, and shock. The author watched this launch from several miles away and still recalls feeling its power. Source: NASA.

Fig. 3.5 The back of the aeroshell for the Mars Science Laboratory Curiosity. The surprisingly large dark circle in the center is the vent (partly blocked by a filter material) to keep the pressure differential during ascent and entry to acceptable levels.
Source: Lockheed Martin.

(the closer an ascent trajectory is to a hover, the more the rocket thrust is wasted fighting against gravity) and is typically accomplished in a few minutes, accompanied by accelerations of a few times Earth gravity (1 $g = 9.81$ m/s^2). The spacecraft structure must be designed to tolerate the associated loads, noting that the load path and orientation on the launch vehicle may have to be quite different from those on the ground or in operation.

Another structural consideration is that the ambient pressure declines from 100 kPa (ie, 10 tons/m^2!) at the surface to zero in space. Any enclosed volume, such as the hull of a vehicle, will suffer a serious pressure differential (enough to deform or burst thin-walled structures) unless it is provided with a venting path of adequate gas conductance. There are standard design rules for such vents (see Fig. 3.5).

This launch depressurization is important to consider electrically: although 1-bar air is a good insulator, as is the vacuum of space, in between a rarified gas can break down, forming an electrical arc.[†] Thus, close attention must be paid to powered electrical conductors that are physically close, and high voltage equipment must typically be powered off during launch and entry.

[†]The curve of breakdown field strength vs pressure is called the Paschen curve. It has a minimum of a few hundred volts per meter for pressures of a few millibar, coincidentally about the pressure at the surface of Mars.

Typically more stressing to a spacecraft designer than the quasi-static pressure and accelerations of launch are the vibration and acoustic loads. The vibration loads are applied (and specified) at some launch vehicle interface and move what the launch vehicle considers its payload. But this payload is the assembly of a cruise stage, an aeroshell, and the landing vehicle itself, which are cantilevered from that attachment interface. The intervening structures—obviously designed to be as lightweight as possible—have natural frequencies at which they may amplify the vibration loads. Thus, the accelerations experienced by systems or components at the extremities of such structures may be much higher than those imposed at the interface. There are usually specifications on the minimum modal frequencies of spacecraft structures (typically a few tens of Hertz). Extensive finite element analysis of the coupled loads and verification by vibration testing are required (see Fig. 3.6). In addition to the structural load path from the attachment interface, some elements can also be vibrated by the intense pressure fluctuations associated with the rocket engines—sound that is not merely at deafening (Fig. 3.7) but actually fatal levels. One danger (and this, at least, is common between spacecraft and helicopters) is that these vibrations can loosen fasteners like nuts, which should be secured with retaining wires.

The attachment of the payload to the launch vehicle is often severed by pyrotechnic devices, essentially contained explosions that can reliably and

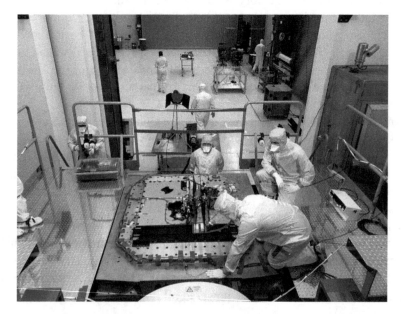

Fig. 3.6 The Mars helicopter and its mounting plate assembly are bolted onto a shake table, which is oscillated by a massive solenoid (visible at bottom as white cylinder) to simulate the vibration of launch.
Source: Lockheed Martin.

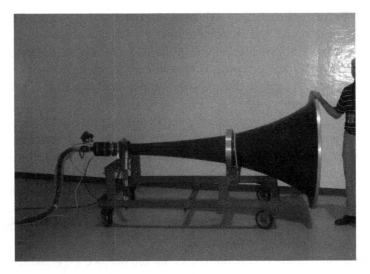

Fig. 3.7 Massive loudspeakers like this one at Goddard Space Flight Center are used to generate and focus acoustic energy to simulate the deafening pressure fluctuations of launch. Source: Author.

instantly cut a bolt. The actuation of these devices causes pyro-shocks, strong but very brief accelerations that can be damaging to very small structures (like wiring). The more massive electronic components often have to be staked onto circuit boards with bracing wire and/or potting material.

ENTRY, DESCENT, AND LANDING (EDL)

The entry and descent to a planetary surface repeats many of the challenges of launch, although for Mars and Titan the entry velocities are typically rather slower (6–8 km/s) than Earth escape. However, this velocity change is accomplished by decelerating aerodynamically inside a capsule or aeroshell. The peak deceleration is generally much larger than those encountered during launch and depends on the entry speed, the area/mass ratio of the aeroshell, the entry angle, and the scale height of the atmosphere (ie, the vertical distance over which the density changes by a factor of $e = 2.718...$).

The Martian scale height at relevant altitudes is about 8 km, whereas for Titan it is some 40 km, due to Titan's low gravity. This results in a very extended atmosphere, which is highly favorable for spacecraft entry.

The kinetic energy of entry must be mostly dissipated into the atmosphere, which happens in the shock layer of flow just in front of the aeroshell. Aeroshells have, counterintuitively, a blunt sphere-cone shape in order to make this shock a strong one and thus to minimize the energy of the flow deposited onto the heat shield itself. In addition to the convective heating of the flow impinging on the shield, there is also radiative heating from the high-temperature gas. This radiative heat loading can be dominant, especially on the shoulder or backshell. Sophisticated computer codes to model the aero-thermodynamics and aerothermochemistry of these conditions, and the response of thermal protection system (TPS) materials (typically phenolic resins filled with silica or cork, or carbon-carbon) to heat loads and shear stress must be measured in plasma wind tunnels that can generate the high-enthalpy flow (see Fig. 3.8).

A shallow entry angle increases the dispersion, because delivery errors get "smeared" downrange, and results in higher total heat loads because the entry takes longer and allows more heat to soak into the heat shield. (Usually the limiting factor is the bondline temperature where the TPS is glued to a sup-porting structure.) However, shallower entries reduce the peak loads, and these are needed on Mars to avoid impacting the ground before the decelera-tion is completed: 6 deg is typical. For Titan, with its much thicker atmo-sphere, entry can be much steeper.

Although all of this is in the domain of the aerosciences, it is ancillary to the rotorcraft, which is protected inside the aeroshell. However, the deceleration, typically 8–15 g for Titan or Mars entries, is a structural design driver (and may necessitate testing in a centrifuge, e.g. Fig. 3.9). Fortunately, this large loading, lasting a couple of tens of seconds, is accompanied by much less vibration than launch. There are, however, pyrotechnic shocks associated with the actuation of mechanisms to release and separate from the heat shield, and

Fig. 3.8 Thermal protection materials for heat shields must be tested in high-energy hypersonic flows, typically generated by heating gas with an electrical arc and expanding in a nozzle (ie, an arcjet). Here, a small capsule model is in an arcjet test, the gas being tortured into incandescence by the flow around the vehicle. The red glow in the shock layer off the conical front shield contrasts with the blue glow in the wake, due to the dif-ferent aerothermochemical conditions.
Source: NASA Photo/Cesar Acosta.

Fig. 3.9 The tolerance of spacecraft to quasi-static acceleration loads is sometimes assessed only by analysis or by direct application of forces to structural elements, but occasionally full-scale accelerations are applied in large centrifuge facilities like this one at Goddard Space Flight Center. Note person in center for scale.
Source: Author.

descent repressurization introduces a pressure differential that must be mitigated by venting, as for launch.

There may be a sudden load pulse associated with the deployment of a parachute (see Fig. 3.10). Because most entry bodies are unstable at flight speeds around Mach 1 (~200–260 m/s at relevant altitudes on Mars or Titan), a drogue or pilot parachute must be deployed before this condition is reached. However, to minimize loads on the parachute (and to avoid aerothermal heating), it should not be deployed at too high a speed—Mach 1.4 is a nominal value. Due to the expense of testing parachute systems via rocket launch or balloon drop, the design of previously flown systems and the flight conditions at which they operated are adopted to the extent possible. Thus, the most common parachute type has a disk-gap-band (DGB) canopy, the gap improving stability and inflation characteristics compared with a circular canopy. DGB chutes have been used on most Mars landers, as well as the Huygens probe.

The parachute system removes the remaining kinetic energy of entry, leaving only that of the terminal (steady-state) descent velocity. The challenge in the thin Mars atmosphere is to do this before running out of sky and hitting the ground; for Titan there is the opposite problem, that the Mach-1.4 condition and chute deployment take place 140 km above the ground, and thus one must

Fig. 3.10 Parachute deployment in the Martian atmosphere, observed by an upward-looking video camera on the Mars-2020 aeroshell. At top left, a pyrotechnic mortar blasts out the parachute in a bag. The lines become taut and the parachute begins to unfurl (top right). The canopy starts to inflate but oscillates (bottom left). Finally, the canopy maintains a stable, fully inflated condition (bottom right). The pattern of stripes allowed any rotation to be measured, but also encoded a message.
Source: NASA/JPL/Caltech.

spend 2 h descending by parachute before landing (with the attendant driving of thermal design and battery energy considerations).

The last step of EDL is, of course, the landing. Some means must be invoked to limit or absorb the kinetic energy. On Mars landings this has been done by some combination of rocket propulsion, shock-absorbing legs, and/or airbags. On Titan, the Huygens probe was not required to survive landing, but the combination of soft ground and the low (5 m/s) 11 mph impact speed allowed it to do so. For Dragonfly, it was realized that the rotor system to be used for later exploration flights was actually the most straightforward and safest means to effect the first soft landing anyway.

PLANETARY SURFACE ENVIRONMENTS

Inside the aeroshell, during interplanetary cruise, the temperature environment can usually be designed to be benign. On the planetary surface, however, the vehicle is exposed to ambient conditions and must be designed to tolerate them.

On Titan, the massive atmosphere warms and cools by only about 1 K, despite the long day–night cycle. The ambient temperature of 94 K is too cold for most mechanisms, and so local electrical heating may be provided to exposed components to warm them (and especially their lubricants) for use. Generally, however, most systems are enclosed within an insulated housing, and electrical or radioisotope heat is applied to maintain a comfortable operating temperature.

On Mars there are wide temperature swings from day to night, because the thin atmosphere retains very little heat. A typical diurnal range at low latitude is 160 K at night to about 270 K at noon. Polar temperatures can reach, and remain at, 150 K, at which temperature carbon dioxide freezes out, forming deposits of frost tens of cm thick over the winter season.

OPERATIONS: COMMUNICATION

The extreme distances of planetary exploration force limits in the amount of information that can be transmitted to or from a vehicle and introduce delays in the process. Radio signals propagate at the speed of light (300,000 km/s), so the two-way light time to the Moon (400,000 km away or quarter of a million miles) is about 2.6 s. It was possible, with some practice and adaptation, for Soviet controllers to joystick a rather slow-moving rover (Lunokhod in 1970) using slow-scan television images for guidance.

Mars varies between 50 and 200 million km from Earth, or a one-way light time of 6 to 24 min. Titan is 72 to 90 light-min away, so real-time control is utterly unfeasible.

If an antenna (Fig. 3.11) such as those of the Deep Space Network (DSN) communicates directly with the vehicle at hand, as in the case of Dragonfly at Titan, then the control response to events could conceivably be a little over 2.5 h (modulo some overhead in decommutating data from telemetry, presenting it in an operations center and then conveying commands from to a DSN station). Data from Titan, however, may only arrive at a few kbps, or perhaps 20 min. to transmit a typical image.

Since the early 2000s, multiple spacecraft have been orbiting around Mars, and these have provided the capability to transfer much more data from surface platforms like landers and rovers than would be possible by communicating directly with Earth. One factor is that orbiters may have relatively large high-gain antennas that can be aimed steadily at Earth and thus achieve high-bandwidth communication. Large antennas are difficult to accommodate on surface platforms (due to EDL geometric constraints), the thermal management challenges of antenna pointing gimbals, and issues such as mobility or wind loads.

The relay infrastructure allows data transfer from platforms that are not in Earth view, because the relay orbiter can be in view of the platform and communicate with it then, and then relay that data to Earth at a later time.

Fig. 3.11 Every pixel of every image we have received from the planets has been delivered as phase, amplitude, or frequency shifts in a radio signal received by radio telescopes like this. (This particular example is actually a Cold War radio telescope in Latvia.) Source: Author.

Depending on the orbit of the relay and the location of the surface platform, there may be an interval of several hours or days between successive relay passes, which typically last 10 or 20 min. The relay passes, using a ultra-high-frequency (UHF) radio link (which allows a relatively wide beamwidth fixed antenna on the surface platform), typically allow the transfer of tens to hundreds of megabits of data.

In the case of Ingenuity on Mars, commands and data are transferred via the Perseverance rover using a short-range radio datalink. Commands to the rover are sent from the DSN, and data from it are sent usually via orbital relay.

In practice, the latency in communication is considerably larger that this, because vehicle commands generally must be embedded in sequences that must be checked and integrated with other data before they are radiated by ground stations of NASA's DSN or the equivalents of other agencies. Furthermore, it is accepted practice that all command sequences be run on an engineering model (typically an electrical replica, but sometimes fully functioning spares) of the vehicle on Earth first, to be sure there are no errors that could cause damage. These command verification steps may take hours or days.

RELIABILITY AND AUTONOMY

Because of the long propagation delays in communication to and from Earth, and the fact that communications are typically only episodic (e.g., when a vehicle is on the side of a planet opposite from the Earth), space systems must have a high degree of autonomy. They not only must perform basic nominal functions like temperature regulation or steady pointing of solar panels towards the sun, but also must be able to tolerate failures without intervention from Earth.

Failure tolerance is typically implemented by redundancy—having backup systems that can execute the function of a failed unit. In some instances this is *hot redundancy* with both units running. In the case of control computers, one will be designated as prime and the other is ready to take over in the event of a failure, or there may even be three in a "voting" arrangement. For other systems, such as solar panels or radio power amplifiers, units may operate continuously in parallel and the system simply operates at lower capability or with less safety margin in the event of a failure. (The same logic applies for twin-engine helicopters.)

Hot redundancy may not be the right approach, however, for life-limited components (e.g., mechanical gyroscopes or lasers) because the life on both units is continuously being used up. There is also a power consumption penalty in running redundant units. Thus *cold spares* are another approach, left switched off until needed. The challenge is then how to reliably detect a failure in the operating unit and activate the spare in a prompt and graceful transition. For units that are in serial strings that must all operate (e.g., the modulator and then the amplifier in a radio transmitter), there is also the architectural question of whether the parallel strings are considered reliable enough or whether the units can be cross-strapped between the strings. Adding systems can actually reduce the overall reliability. An entire field of reliability engineering considers such questions.

The considerations of fault protection and reliability thus form a major element of space systems engineering.

REFERENCES

1   For a comprehensive examination of such lessons, see Harland, D., and Lorenz, R., *Space Systems Failures*, Springer-Praxis, 2005.
2   There are many excellent texts on spacecraft design, all with their own strengths: Larson, W. J., and Wertz, J. R., *Space Mission Analysis and Design*, Microcosm, 1992; Brown, C. D., *Elements of Spacecraft Design*, AIAA, 2002. Griffin, M. D., *Space Vehicle Design*, AIAA, 2004. The astronautics syllabus that I covered as an undergraduate is largely covered in Fortescue, P., Swinerd, G., and Stark, J. (eds.), *Spacecraft Systems Engineering*, John Wiley & Sons, 2011. A focus on some different aspects of space engineering is in Cruise, A. M., Bowles, J. A., Patrick, T. J., and Goodall, C. V., *Principles of Space Instrument Design*, Cambridge University Press, 2006. For details on the specific problems of engineering for entry probes and landers, see Ball, A. J., Garry, J. R. C., Lorenz, R. D., and Kerzhanovich, V. V., *Planetary Landers and Entry Probes*, Cambridge University Press, May 2007.

Chapter 4

INGENUITY: DESIGN OF THE MARS HELICOPTER

"The helicopter is probably the most versatile instrument ever invented by man. It approaches closer than any other to fulfillment of mankind's ancient dreams of the flying horse and the magic carpet."

—Igor Sikorsky

As a technology demonstration, the goals of the Mars helicopter might be succinctly if informally described as "fly on Mars, and take a picture." Perhaps analogous to Kennedy's Apollo specification of "Man, Moon, Decade," it is tempting to speculate that the initial concept was of a 1-kg vehicle able to fly for 1 min., although the only support for this is the mention of 1 kg (2.2 lbs) in the first public report on the helicopter effort. (A similar logic led to my own proposal six years earlier of the 1-kg Titan Bumblebee.)

Even a 1-kg vehicle demands high performance to claw enough lift in the thin Mars atmosphere, and the resultant Mars helicopter design is unsurprisingly dominated by rotors that are as large as feasible to keep the disk loading down. Delivery to the Mars surface as an add-on element on the already-crowded Mars 2020 Perseverance rover required a compact deployable configuration, hence the coaxial design with two rotors, each with two blades (see Fig. 4.1).

As the various design aspects required to support standalone operation on the Mars surface, and to tolerate the wide temperature swings there, were refined, and the formidable challenges of stability and control were grappled with, the vehicle mass grew by some 80% to 1.8 kg (4 lbs). Another factor is that the launch vehicle vibration loads (7.9 g rms [root mean squared load]) and pyrotechnic shock loading drove the helicopter structure to be stiffer than would be required for Mars operation alone. Nonetheless, the Mars helicopter represents a triumph of miniaturized design and construction and has performed admirably, as described in Chapter 6.

Fig. 4.1   The flight model of the Mars helicopter. Note the "horns" near the roots of the rotor blades (Chinese weights) and the black/reflective texture of the electronics box.
Source: NASA/Jet Propulsion Laboratory (JPL).

One aspect of the helicopter that was perhaps not appreciated early in the design is that the low density of the Mars atmosphere (assumed at its planned operating location to be in the range of 0.014–0.02 kg/m^3) not only makes it much harder to generate lift, but also means there is essentially no aerodynamic damping of the blade motions. This lack of damping allows the flapping mode of the rotor blades to couple into the body of the helicopter and cause the system to become dynamically unstable.* As will be seen, the thin atmosphere is also unhelpful in that it does little to mitigate heat buildup in the propulsion motors. The material in this chapter is drawn principally from the papers by Balaram et al.[1] and Pipenberg et al.[2].

ROTOR SYSTEM DESIGN

The rotor system consists of the rotor blades and hubs, swashplates and control linkages, servo actuators and feedback system, propulsion motors and associated power electronics, main mast structure, and wiring harness. The upper rotor is an almost identical copy of the lower rotor with the exception of the rotation direction of the rotor blades. The direct-drive brushless propulsion motors are mounted to the rotor hubs.

ROTOR BLADES

The Mars helicopter has 1.21-m-diameter (47.6 inches) rigid rotors (ie, no flap or lag hinges). The rotors are challenged by a very low operating Reynolds number (below 11,000) and relatively high tip Mach number

---

* The low atmospheric density reduced aerodynamic control authority and damping relative to the blade inertia, characterized by a much lower Lock number than is typical for terrestrial helicopter rotors.

(~0.7) at their peak rotation rate of 2800 rpm. These aerodynamic aspects would lead one to favor a thin airfoil to minimize drag and thus the power needed for flight; however, the need to provide adequate structural stiffness instead pushed the blade design to have a deeper section. The rotor blade design was the result of extensive computer simulation as well as wind tunnel and structural testing, building on the wide experience of AeroVironment Inc. in high-altitude flight and lightweight propellers such as those of the human-powered Gossamer Albatross. The adopted airfoil (modified CLF5605) had a peak camber aft of the midchord to delay flow separation and maximize the lift coefficient. The blade was tapered strongly (see Fig. 4.2) to minimize the mass near the tip, which has a strong effect on the first flap mode frequency.

This rotor blade design was shaped to blend the outboard aerodynamic surfaces into a cylindrical tube root design that could interface to the pitch bearings and hub. The blades feature a super-lightweight composite construction with pre-impregnated carbon fiber (prepreg) molded over machined polymethacrylimide structural foam (Rohacell) cores. High-strength unidirectional carbon fiber was used to create stiff spars, and bidirectional woven carbon fiber with low areal mass (80 g/m^2) was used for the skins. This fabrication results in blades with a remarkably low mass, less than 33 g (1.16 oz) each.

Spacecraft contamination control requirements (recall that Perseverance is to encapsulate samples of Mars rocks for return to Earth) require that polymer materials have low outgassing. A cyanate ester resin was used in the composites to minimize water vapor retention and the outgassing rate and to avoid degradation in the harsh ultraviolet flux on Mars. The components in the rotor system were fabricated and assembled in a cleanroom and were baked out in a vacuum chamber, also to reduce outgassing.

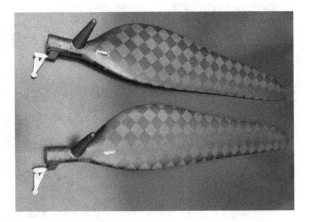

Fig. 4.2   Ingenuity's rotor blades, about 60 cm long. The chequerboard pattern is due to the interleaving of carbon fibers in the blade skin. The conical projections at left are Chinese weights, and the metal parts at far left are the levers through which the swashplate controls the blade pitch.
Source: AeroVironment, courtesy B. Pipenberg.

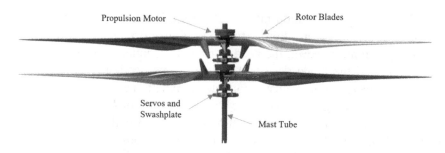

Fig. 4.3 Side view of rotor system.
Source: AeroVironment, courtesy B. Pipenberg.

ROTOR HUB AND PROPULSION MOTORS

Even though the rotor blades are light, the high rotation rate causes strong loads to be applied to the hub. The maximum centrifugal force is approximately 550 N (125 lb), and worst-case hub moments are approximately 7.5 Nm (65 in.-lb.) per blade. Thus the hub structure must be stiff: The carbon fiber tube acts as the main structure of the helicopter. The 81-conductor wiring harness from the main helicopter electronics to the actuators and to the solar panel and antenna pass through this tube (see Fig. 4.3).

The blades mount to the hub via ceramic (silicon nitride) bearings and are spun by direct-drive brushless outrunner motors that were custom-designed and built at AeroVironment. The stator is fabricated from laser-cut laminations and wound with copper wire of rectangular cross-section to maximize the winding density. The 46-pole magnet ring is bonded directly to a motor bell mounted on the hub. The high pole count and copper density gave a motor efficiency of ~80%.

Fig. 4.4   Ingenuity's propulsion motor. Left: The motor windings use square-section wire to maximize the copper packing density. The beryllium heat sink is mounted to the central structural shaft. Right: The outer, rotating carbon fiber structure to which the blades attach. Note the ring of little rectangular magnets. The development of powerful and compact rare-earth magnets in the last quarter century has been an important enabling technology in the drone revolution.
Source: AeroVironment, courtesy Ben Pipenberg.

The central motor hub is cut from a beryllium alloy AlBeMet AM162 (see Fig. 4.4). Beryllium offers outstanding stiffness for low weight but is usually avoided in manufacturing because of the difficulties of working with it. (Machining it yields toxic dust.) However, here the material was necessary not just because of its strength and light weight, but also because of beryllium's high heat capacity, allowing this part to act as a heat sink during flight operations. Even with this protective measure, however, the motor warms by about 1°C/s (1.8°F/s), and motor overheating is, in fact, the limiting factor on flight duration. Again, the thin Mars atmosphere introduced a significant challenge only evident once detailed design and testing were performed—the low density makes convective cooling relatively ineffective.

Flight Control System

Propulsive power is useless without control. To effect flight in a desired path, and to handle wind gusts up to 9 m/s (20 mph) during operation, the Mars helicopter design incorporates collective and cyclic control on both the upper and lower rotors. Collective changes the pitch of the blades as a pair to control thrust of the rotor; cyclic changes blade pitch as a function of the rotor position (ie, cyclically) to provide lateral control. Yaw control is achieved with differential collective pitch between the upper and lower rotor to produce a net yawing moment without changing the net thrust on the vehicle. (The rotation rates of the two rotors are kept constant.) There are swashplates, machined from titanium alloy, for both the upper and lower rotor, each with collective and cyclic control (see Fig. 4.5). The collective angle can range from –4.5 deg to 17.5 deg, and the cyclic angle has a range of 10 deg. The accuracy of these deflections was required to be within 0.25 deg of the commanded position.

Mechanical control inputs to the rotor system are provided by servo actuators composed of three Maxon brushed DC motors (DCX10). These motors

Fig. 4.5 Left: Swashplate assembly, with ball linkages shown. Right: Servo motor arrangement.
Source: AeroVironment, courtesy Ben Pipenberg.

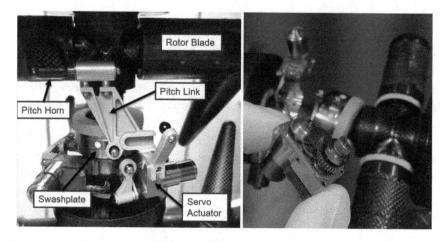

Fig. 4.6 Left: Rotor hub assembly showing the blade pitch horns linked to the swashplate via ball couplings. Right: The exposed gearbox of one of the swashplate servo actuators. Source: AeroVironment, courtesy Ben Pipenberg.

are mounted in an aluminum housing and use Hall sensor magnetic feedback control to achieve a bandwidth of better than 12 Hz. They act through a five-stage gearbox of brass, steel, and hard-anodized aluminum gears. The gear ratio was selected to shift the torque/speed curve such that the worst-case operating point (900 deg/s at 250 mNm) could still be achieved at the minimum operating voltage of 16 V.

The servos control the height and tilt of each swashplate through pitch links (Fig. 4.6) machined from polyether ether ketone (PEEK). These links are locked during launch and transit to Mars to prevent damage to the servos due to launch vibration. Resistive heaters are located on the magnetic encoders to preheat the motor components prior to operation (rather like a bumblebee warming its muscles before flight).

Tungsten carbide weights are molded into the rotor blade roots. These Chinese weights provide a restoring force on the blade moments when under centrifugal loads, thereby reducing the torque requirements on the swashplate actuators.

The internal volume of the swashplate assembly must be vented to prevent damage during the rapid ascent depressurization during launch. However, the vent must guard against intrusion of dust from the Martian atmosphere; thus, there are labyrinth seals around the bearings and a fiber high-efficiency particulate air (HEPA) filter.

SENSORS

Control is impossible without sensing. The vehicle must have knowledge of its orientation and position with respect to the ground, which it estimates by fusing together data from several different sensors. Two decades ago these elements were massive and expensive, but today the sensors are mass-produced and miniature devices, and the data processing is readily supported by modern processors.

The lowest-level element of this guidance and navigation is the orientation, which is measured with inertial measurement units (IMUs). An IMU combines rotation measurements [historically, spinning gyroscopes, but here solid-state microelectromechanical systems (MEMS) silicon rotation sensors] with acceleration measurements to estimate position and orientation changes from some initial state. The Mars helicopter uses two Bosch Sensortec BMI-160 IMUs, one for the upper sensor assembly in a vibration isolation mount and one for the lower sensor assembly where it is colocated with the cameras. These report three-axis accelerations at 1600 Hz and angular rates at 3200 Hz.

Accelerometers tend to have a small zero-offset or bias term. When the navigation system integrates this bias it leads to a position error that grows quadratically with time, and thus additional sensing of position or velocity is essential to rein in the divergence of the solution. First, a two-axis MEMS MuRata device (SCA100T-D02) is used to estimate the accelerometer bias before takeoff. Even with this correction, a small MEMS IMU would be far too inaccurate to control landing (and the terrain might be uneven, so altitude relative to the takeoff site might be different from that relative to the landing site), so local ground-relative altitude is measured with a laser time-of-flight altimeter (Garmin Lidar-Lite-V3) with a range of tens of meters at 50 Hz.

Similarly, the inertial horizontal position estimate would diverge unacceptably without correction. On many terrestrial drones or other vehicles this correction is provided by Global Positioning System (GPS) satellite signals, a capability now often taken for granted in every smartphone. However, there is no GPS on Mars. On planetary landers like Viking, the horizontal position/velocity measurement for safe landing was provided by Doppler radar, but radars are typically rather bulky and massive. The technology of the last decade or two now allows horizontal navigation to be performed optically, with a small camera. (The camera, of course, is actually the easy part—it is the feature extraction and correlation operations that must be performed on the images, and the subsequent geometric transformations to determine the navigation information,[†] that are novel!)

Thus, the Mars helicopter has a small down-looking navigation (NAV) camera. This is a global-shutter, grayscale 640×480 pixel (VGA) detector, the Omnivision OV7251. Its optics give a field of view (FOV) of 133×100 deg centered on nadir with an average instantaneous field of view (IFOV) of 3.6 mRad/pixel, and is capable of acquiring images at 30 frames/s. Under typical Martian surface illumination, the exposure times are expected to be a few

---

[†]These functions were first executed on Mars by the Descent Image Motion Estimation System (DIMES) cameras that actuated side-firing rockets on the Mars exploration rovers Spirit and Opportunity in 2004 to null out horizontal velocity due to wind just before landing.

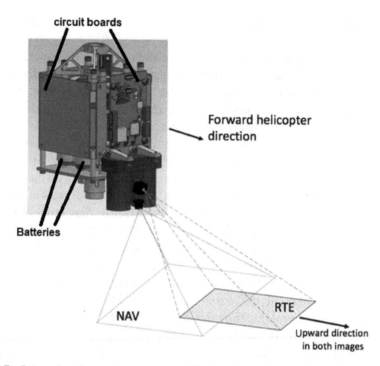

circuit boards

Forward helicopter
direction

Batteries

NAV

RTE

Upward direction
in both images

Fig. 4.7   Internal equipment arrangement. The circuit boards are arranged in a box shape around the batteries. The RTE and NAV cameras look in a down/forward direction.‡
Source: NASA.

milliseconds. Visual features are extracted from the images and tracked from frame to frame to provide a velocity estimate. The optics are protected by a small window.

In addition to the NAV camera, there is a payload called the return-to-earth (RTE) camera (see Fig. 4.7). This is a rolling-shutter, high-resolution 4208×3120 pixel sensor (Sony IMX 214) with a Bayer color filter array mated with an O-film optics module. This camera has an FOV of 47 deg (horizontal)×47 deg (vertical) with an average IFOV of 0.26 mRad/pixel. The RTE camera is pointed approximately 22 deg below the horizon, resulting in an overlap region between it and the NAV camera's FOV of approximately 30 deg.

AVIONICS

The flight control and higher functions of the helicopter are effected by three processors. Two flight control units are automotive-grade microcontrollers; the

---

‡The RTE camera is shown here in a portrait orientation, as originally intended. In fact, during assembly, it was found that a cable bundle prevented this mounting, and so the camera had to be rotated. Thus the RTE images from Mars in Chapter 6 are in landscape orientation, and almost never show the horizon. I am grateful to Bob Balaram for explaining this.

higher-level functions and image processing are performed by a Qualcomm Snapdragon processor, comparable with that in a modern smartphone.

The Snapdragon 801 has a 2.26-GHz quad-core processor with 2 GB random access memory (RAM), 32 GB flash memory, and various interface ports. This processor performs visual navigation via a velocity estimate derived from features tracked in the NAV camera and uses this velocity estimate in its navigation filter. Aside from these flight control functions, the Snapdragon effects command processing, telemetry generation, and radio communication.

The two flight-control (FC) microcontroller units (MCUs) are Cortex R5 devices. They communicate with the Snapdragon via a universal asynchronous receiver/transmitter (UART). These FC processor units operate redundantly, performing the high-speed inner control loop of the flight actuators. At any given time, one of the MCUs is active with the other waiting to be hot-swapped in case of a fault.

The avionics are mounted on five printed circuit boards that form the five facets of the electronic core module (ECM) cube, enclosing the battery pack (see Fig. 4.8). The bottom of the cube is the battery interface board (BIB), which is attached to the battery and hosts the battery monitoring circuitry, motor power switches, and motor current monitors. The battery and the BIB can be removed and replaced without dismantling the entire helicopter.

The remaining four boards are the field programmable gate array (FPGA)/ flight controller board (FFB); the NAV/servo controller board (NSB), which hosts the Snapdragon; the telecom board (TCB); and the helicopter power board (HPB). The FFB hosts the two FC processors, operating in sync with the

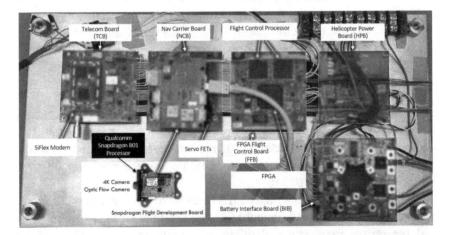

Fig. 4.8 "Flatsat" bench checkout arrangement of an early prototype of the helicopter electronics. The boards form the faces of a box, but here are unfolded flat for testing. This test setup has a number of nonflight parts, cables, and connectors but the overall size and arrangement of the circuit boards is representative. Note that the image has been selectively blurred to avoid proprietary or export control issues.
Source: JPL.

same clock and data provided by an FPGA chip, which handles the interface to all the sensors and actuators. The FPGA is the heart of the helicopter avionics and implements the custom digital functions not implemented in software due to speed or hardware limitations of the processors. The FPGA is a military-grade version of MicroSemi's ProASIC3L.

The FPGA performs vehicle flight control including an attitude control loop operating at 500 Hz. It performs motor control, generating pulse width modulation (PWM) signals for the control servos and commutating the currents to the brushless rotor propulsion motors. It turns on and off the other avionics elements as they are needed, monitors temperatures and switches heaters on as needed, monitors the battery cell voltages, and performs cell balancing. Being always-on, the FPGA maintains spacecraft time.

The allowance of commercial off-the-shelf (COTS) electronics parts on a NASA Class D technology demonstration permits these high-performance processors and sensors to be used. The range of parts that are space-qualified and allowed to be used on regular NASA projects is much narrower, and parts typically are physically larger and/or need higher power. The COTS parts used on the helicopter are, however, not consumer-grade parts, but have a military or industrial extended operating temperature range of at least –40°C to +85°C (–40°F to 185°F). Although radiation on the surface of Mars is less severe than in space, radiation precautions were nonetheless taken in the electronics design. The FPGA has triple-module redundancy to protect against single-event upsets (SEUs, where a cosmic ray flips a memory bit) and is operated at a reduced speed to minimize the effect of signal propagation delays due to the total radiation dose. Single-event latchups (SELs) are detected with current monitors and can be cleared by power cycling. In addition, the helicopter avionics have current limiting to prevent the most destructive SEL events, and most devices are switched off when not in use to minimize their exposure to SELs.

TELECOMMUNICATION SYSTEM

The Mars helicopter does not communicate directly with Earth, but uses the Perseverance rover as a relay. An ultra-high-frequency (UHF) radio link is designed to relay data at rates of 20 kbps or 250 kbps over distances of up to 1000 m. (Compare these rates with a 1990s-era telephone modem of 56 kbps or modern WiFi of 10 Mbps.)

The link is implemented with a COTS Institute of Electrical and Electronics Engineers (IEEE) 802.15.4 standard (Zig-Bee) digital radio, a system widely used for sensor networks and other low-bandwidth links. Operating at 900 MHz, the radios (one as the helicopter base station on the Perseverance rover, the other on the helicopter) use a SiFlex 02 chipset, originally manufactured by LS Research. The radio emits approximately 0.75 W of radiofrequency (RF) power at 900 MHz with the board consuming up to 3 W of DC power when transmitting and ~0.15 W in receive-only mode.

The helicopter antenna is a loaded quarter-wave monopole mounted near the center of the solar panel (which acts as ground plane) at the top of the entire helicopter assembly and is fed through a miniature coaxial cable routed through the mast.

One difficulty in using COTS electronics systems to be used on Mars is the very wide swing in temperatures. Not only can the deep thermal cycling cause stresses from differential expansion that can fatigue and break wires or solder joints, but if the transmitter and receiver are at very different temperatures, the radios can be detuned from each other due to the temperature dependence of their oscillators. This was, in fact, a severe problem for the COTS radio used on a previous technology demonstration, the Sojourner rover on Mars Pathfinder in 1997. Although the electronics on the helicopter are kept warm (above −15°C/5°F) by heaters to avoid deep chilling, the antenna and cable are exposed to temperatures as low as −140°C (−220°F) at night.

POWER AND THERMAL SYSTEMS

The helicopter is powered by a battery recharged with a solar panel. At Perseverance's location and initial season, the battery could recharge completely in a single sol (Martian day). The solar panel is mounted on the mast tube on the top of the helicopter (see Fig. 4.9); mounting at the center minimizes the aerodynamic obstruction of the flow induced by the rotors. The panel is not circular or square, but rather is rectangular to provide a large enough energy-generating area (544 cm^2 of active cell area on a 680 cm^2 substrate) while still fitting in the geometric envelope allowed by the cramped accommodation on the rover belly. The solar panel is covered in multijunction inverted metamorphic (IMM4J) cells

Fig. 4.9   Technicians in bunny suits inspect the Ingenuity helicopter. This view shows the checkerboard pattern of the bidirectional carbon fiber weave of the rotor blades and the arrangement of solar cells on the panel above.
Source: NASA/JPL.

made by SolAero. These cells are optimized for the Mars solar spectrum (where the dust absorbs blue light and scatters red light).

The helicopter battery comprises six lithium-ion cells (Sony SE US1865o VTC4) with a nameplate capacity of 2 Ah. The mass of the six cells is 273 g. The FPGA attempts to balance the charging of the cells to provide a uniform cell voltage (4.25 V maximum). The battery operating voltage is between 15 V and 25.2 V, depending on charge state.

The margined end-of-life battery capacity at $0°C$ ($32°F$) is determined to be 35.75 W-h. Most of this energy is budgeted for night-time survival heater operation (estimated at 21 W-h for typical northern spring operation). Approximately 10 W-h is available for flight of about 90 s at a steady 360 W with a short burst of higher-power operation up to 510 W. (The cells are rated for a maximum discharge current of 25 A.) Thirty percent of the battery capacity is considered reserve. Note that battery capacity is somewhat improved at modestly higher temperatures, so energy performance may be slightly better than this. [In fact, the cells are warmed to 20 deg C (68 F) to permit high-current operation.] The battery is the most thermally sensitive component, and Kapton film heaters on the cells are operated to maintain their temperature above $-15°C$ ($5°F$) at night.[§]

To minimize the energy expenditure needed to maintain benign battery and electronics temperatures, they are kept inside the helicopter warm electronics box (HWEB). This enclosure provides insulation via aerogel and a 3-cm gap of carbon dioxide (Mars atmosphere); the outer surface is a Sheldahl polyamide film with high solar absorptivity (0.8) and low thermal infrared emissivity (0.1).

A challenge for insulation of spacecraft and instruments is that the insulation must have penetrations for various functions—for example, there have to be windows in the bottom of the HWEB for the laser altimeter and cameras. Furthermore, conductive losses occur through the wiring harness—the wire diameters are selected to be of the thinnest gauges that can still support the current flow during operations without excessive heating or voltage drop.

During flights, the electronics and battery warm up as a result of avionics operations (most particularly the Snapdragon processor) and battery self-heating. However, the dissipation is modest enough and the thermal inertia high enough that there is no overheating during the short flights. Of course, this is not the case for the propulsion motors discussed earlier.

LANDING SYSTEM

The landing system (see Fig. 4.10) keeps the body of the helicopter off the ground to tolerate some surface roughness and provides energy absorption at

---

[§]Teddy Tzanetos notes that on the night before a planned flight, the battery temperature is allowed to drop to -20°C, to save some energy.

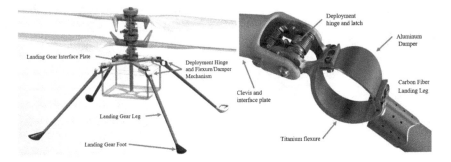

Fig. 4.10   Landing gear arrangement and closeup of the hinge/damper assembly.
Source: AeroVironment, courtesy B. Pipenberg.

landing. It is designed to permit landing on a surface with slopes up to 10 deg with the vehicle at a roll (or pitch) angle of 30 deg. Vertical velocity at the height where the passive gravity drop is initiated can be as high as 2.5 m/s. A horizontal velocity of up to 0.5 m/s can be present due to delivery errors in the control system. The large footprint with a side length of 577 mm (22.7 in.) provides a stable base for the helicopter and reduces risk of tip-over.

Rigid, lightweight legs with high resonant frequencies are necessary to avoid any interaction of landing gear structural modes with the control system. The four legs are tapered carbon fiber/epoxy tubes. The feet are designed to prevent the legs from digging into soft landing surfaces. They also provide some damping by dragging against the ground as the legs bend on titanium flexure hinges [deflecting as much as 15 deg to provide an effective vertical stroke of 92 mm (3.62 in.) for 2 m/s impacts]; additional energy absorption is provided by a yielding 1100-series aluminum alloy damper.

Latching deployment hinges at the top of the legs allow the landing gear system to be folded and released from a stowed position on the belly of the Perseverance rover. Two of the legs deploy passively as the aircraft is rotated away from the rover belly pan while the other two deploy when a pair of restraints are released. After deployment, spring-loaded pawls lock each leg in place.

REFERENCES

1   Balaram, B., Canham, T., Duncan, C., Grip, H. F., Johnson, W., Maki, J., Quon, A., Stern, R., and Zhu, D., "Mars Helicopter Technology Demonstrator," *2018 AIAA Atmospheric Flight Mechanics Conference*, p. 23.

2   Pipenberg, B. T., Keennon, M., Tyler, J., Hibbs, B., Langberg, S., Balaram, J., Grip, H. F., and Pempejian, J., "Design and Fabrication of the Mars Helicopter Rotor, Airframe, and Landing Gear Systems," *AIAA Scitech 2019 Forum*, p. 620.

Chapter 5

TESTING AND DELIVERING INGENUITY

"In the course of your work, you will from time to time encounter the situation where the facts and the theory do not coincide. In such circumstances, young gentlemen, it is my earnest advice to respect the facts."

—Igor Sikorsky

The development effort that led to Ingenuity's flight began with preliminary studies and rotor tests by the Jet Propulsion Laboratory (JPL) and AeroVironment in 2013 and 2014. Balaram's report[1] began promisingly:

We describe successful tests of a Mars rotor airfoil that were recently conducted in the JPL 10-foot pressure chamber under Mars ambient conditions. The airfoil was designed and fabricated at AeroVironment Inc. Torque and thrust data were collected over a range of rotor RPMs, chamber pressures and airfoil blade angles. Test results confirm that vertical take-off and landing flight on Mars is feasible, and that the lift performance closely matches predicted analytical CFD results.

When a free-flight test vehicle with representative rotors was first tried in Mars pressure conditions in a chamber at JPL in 2014, the results showed that (just like the earliest days of terrestrial rotorcraft) controllability was an unexpectedly difficult challenge. At this early stage, the test vehicle just had the rotors, motors, and swashplate mounted on some landing gear. Power and control signals from a joystick were conveyed by a cable. Not having all the sensors, battery, and structure allowed the "skeleton" vehicle to weigh approximately what the fully equipped system would weigh on Mars.

Balaram's report gives the merest inkling of what happened: "The resulting airfoil design can be used as part of a dual co-axial rotorcraft vehicle architecture which we describe together with some of the associated flight control issues." Video released years later showed that, indeed, it flew, but within seconds it slammed into the walls of the test chamber and its rotors smashed to pieces,[2] despite the human pilot having considerable experience with experimental small-scale aircraft. AeroVironment's Matt Keennon noted in an interview, "It's inherently unstable. . . . We've tried flying this type of helicopter without a control system. It goes bad quickly."[3]

Part of the challenge was the weak damping in the low-density atmosphere mentioned in Chapter 2. Another issue is that the relatively high rotation rate gives strong gyroscopic effects on the rotors themselves. (Because there are two rotors rotating in opposite directions, such that their angular momenta are virtually equal but of opposite sign, the net angular momentum and thus gyroscopic stability of the vehicle as a whole is small.) The tendency of the blades, which had much more twist than is typical for a helicopter, was to try to rotate to flatten[4] themselves in the spin plane—the so-called "tennis-racquet effect."

Stability augmentation by the electronic control system would be essential. As the design matured, tests were performed for system identification,[5] to understand all the couplings between control inputs and dynamical effects. Some of these used one of two engineering development models (EDM; EDM1 was the flight test workhorse, while EDM2 was mainly used for thermal-vacuum testing) of the helicopter held securely in a gimbal, allowing the roll and pitch damping to be measured. In these tests, the helicopter was in fact mounted upside down, because the dimensions of the test chamber allowed the downwash from the rotors to circulate better to keep them out of ground effect.

Other tests were performed with the helicopter on a swinging arm (Fig. 5.1). (In fact, a swinging arm apparatus was used by English engineer John Smeaton in 1759 to measure the aerodynamic forces on windmill blades.[6]) This allowed an edgewise flow to be imposed on the rotors, so that the effect of horizontal flight on stability could be measured. These effects* were significantly larger than expected. Wind tunnel tets (Fig. 5.2) and Computational Fluid Dynamics simulations (e.g., Fig. 5.3) show the complexity of coaxial rotor flowfields in the presence of cross-flow.

The other major stability effect was that the roll and pitch damping coefficients had a positive sign, which is not typical helicopter behavior. This meant that roll and pitch rates were self-amplifying, rather than damped, and resulted in part from the flapping dynamics of the blades.

As these stability effects became understood and incorporated into the flight control system (the control loop on the rotors would operate 500 times per second) and the rotor blade design evolved to have increased stiffness, the flyability of the vehicle improved.

---

*manifested in the derivatives of the moment coefficients with respect to speed, a couple of terms in the plant matrix.

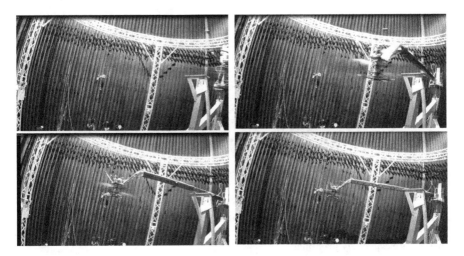

Fig. 5.1 Ingenuity development model mounted on a swinging arm inside the 85-ft chamber at Mars pressure. By swinging the helicopter sideways, a controlled cross-flow can be induced on the rotors to determine the dynamic response.
Source: NASA/JPL/Caltech.

Fig. 5.2 A rotor test at Mars pressure in a wind tunnel at the NASA Ames Research Center. This test rig performed the first coaxial rotor test in cross-flow at Mars conditions.
Source: NASA.

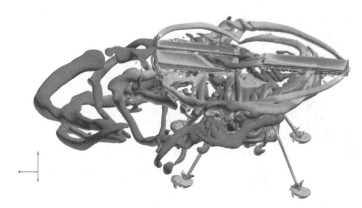

Fig. 5.3   Computational fluid dynamics simulation showing the vorticity in the flow produced by Ingenuity's contrarotating rotor blades. The rotor leading edges and the tip vortices are especially prominent.
Source: Courtesy Daniel Escobar and Anubhav Datta.[9]

The first full-size vehicle (see Fig. 5.4) flew in a representative Mars environment in 2016. The 25-ft space simulation chamber at JPL, some 85 ft tall and more typically used to do thermal-vacuum testing on large spacecraft like Cassini, was used. It takes some 2 h to pump the air out of this massive volume. Carbon dioxide was then used to back-fill the chamber to Mars pressure,

Fig. 5.4   The first full-scale prototype Mars helicopter. Notice that the equipment/battery box is not present beneath the rotors—power and control in this test were supplied by the cable. Notice also the set of small grey spheres above and below the rotors. These are reference markers to allow a Vicon video tracking system to measure the position and rotation of the vehicle. The arrangement is deliberately asymmetric to avoid ambiguities in the estimated orientation.[†]
Source: NASA/JPL/Caltech.

[†]This vehicle was referred to by the team as the 'Phase 3' vehicle : it had cyclic controls only on the lower rotor and collective only on the upper one, and both rotors were driven by a single motor via gears. The EDMs had cyclic control on both rotors and individual direct drive motors for each rotor.

yielding a density of 0.0175 kg/m^3. The flight was fully autonomous, consisting of takeoff, climb to an altitude of 2 m at a rate of 1 m/s, hover for 30 s, descent at 0.5 m/s, and landing. This time the vehicle (Fig. 5.22) flew steadily.

A higher-fidelity helicopter model, EDM-2, was later built. For tests where the vehicle was secured to a test stand, a 6-degree-of-freedom force/torque sensor was used to measure the aerodynamic effects. For free-flight testing, a Vicon motion tracking system with 18 cameras was used to triangulate the orientation and position history of the vehicle to deduce its dynamics (Fig. 5.5). Free-flight testing with the all-up vehicle (which weighs 2.5 times what it will weigh on Mars, due to Earth's higher gravity) was performed by suspending the rotorcraft from a gravity offload system. This maintained a constant tension on the support line to reduce the effective weight. In other vehicle tests (see "Thermal Testing" section) temperature sensors and a thermal camera were used. Stroboscopic lighting and vibration-monitoring accelerometers informed the structural dynamics of the vehicle. Other tests (Fig. 5.6) tested the navigation cameras in powered flight, to be sure rotor vibration did not cause image blur or other problems.

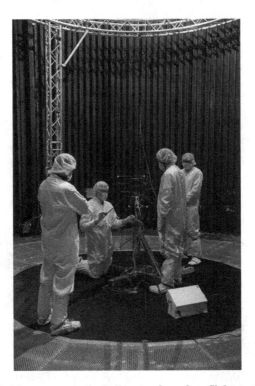

Fig. 5.5 JPL technicians prepare the helicopter for a free-flight test in the 25-ft space simulation chamber. The Vicon video tracking markers are visible above and below the body of the helicopter.
Source: NASA/JPL/Caltech.

Fig. 5.6   Free-flight hover test in the 85-ft chamber, exercising the navigation camera capability with the rigors of flight vibration. Left: A large number of artificial features for the camera to lock on to have been created on the floor with adhesive tape. Right: The downward-looking camera view from the helicopter.
Source: NASA/JPL/Caltech.

WIND WALL

Response to winds and higher-speed forward-flight conditions than could be generated with the swinging arm was measured by placing the rotorcraft in front of a wind wall, which consists of an array of small fans and is capable of producing winds up to approximately 10 m/s (see Fig. 5.7). One reason for using an array of small fans instead of one big propeller as in most traditional wind tunnels is that modern computer control can adjust the relative speeds

Fig. 5.7   Ingenuity EDM in front of the wind wall inside the 85-ft JPL chamber. The vehicle is mounted on a gimbal to study the effect of cross-flow in a variety of orientations. Note the vehicle is upside-down in order to minimize ground effect. Also note the many bundles of wires at the right, supplying power and control to the fans that make the black wall.
Source: Image courtesy Jason Rabinovitch, from [8].

of fans across the array to develop a more uniform flow, regardless of the effects of the chamber walls. It is notably difficult to operate a wind tunnel stably at low speeds. A wind wall is also scalable; the system can be developed and proven out with an array of 16 or 25 fans, and then more fans added for a wider test section. An important factor for testing of small aerial vehicles is that each small fan responds quickly to a change in demanded speed, whereas a large propeller may take several seconds to spin up or spin down. Thus, to simulate gusts with rapid changes in wind speed, a wind wall is a much more effective setup.

Although the wind wall used in the JPL test was not itself flight hardware, because it was installed in the space simulation chamber, it had to be subject to the same outgassing rules. Thus, the standard polyvinyl chloride (PVC) insulation on the wiring to the fans had to be replaced with Teflon to avoid depositing excessive contaminants in the chamber.[7] The number of fans (441) was such that cabling to each individually would be prohibitive, so a local controller was used on each 3×3 module. To generate a 10 m/s flow, some 3.7 kW of power had to be fed into the chamber. In the low-pressure Mars tests, the convective cooling of the fan motors was much less effective, and so tests were limited to 10 min in duration to avoid overheating.[8]

LANDING GEAR TESTING

One of the virtues of a small vehicle is that it can be easy to test at full scale. Thus, rather than relying on computer models to analyze the landing dynamics and tune the stiffness and damping of the landing gear, it was decided simply perform physical tests with a 1:1 model (see Fig. 5.8).[10]

The test setup used a gravity offset winch system that was attached to a surrogate test article (a weighted box without electronics, rotors, and the like, because landings would be performed on Mars with the rotors being powered

Fig. 5.8   Tests of the Ingenuity landing gear at AeroVironment. A sliding rail supported by ladders provides a controlled vertical and horizontal velocity of the impact on various simulated planetary surfaces. The suspension lines off-load the test article's weight to simulate Mars gravity.
Source: AeroVironment, courtesy Sara Langberg.

down). The winch was attached to a motorized horizontal gantry to simulate side velocities of up to 0.5 m/s, and a release device was built to allow the vehicle to be dropped with consistent initial conditions of rotation and velocity. The impact conditions and the bounce response were captured with a Vicon video camera system. The landing surface of quartz sand, or coarse-finished concrete to simulate Mars bedrock, was tilted up to 10 deg from horizontal. Rocks up to 5 cm high were arranged to test interaction with sharp-cornered surface features.

THERMAL TESTING

Tests in the JPL 25-ft chamber were important in refining the thermal design of the helicopter.[11] The thermal balance, in terms of how fast heat would leak out of the system to the environment, would be vital in ensuring that the amount of electrical heating needed at night would not exceed the energy budget provided by the solar panels during the day. This, in turn, depended on conductive leaks through structural materials and wires, and on radiative heat transfer and gas convection from surfaces (Fig. 5.9). The inner walls of the space simulation chamber were chilled with liquid nitrogen to replicate the cold sky radiance on Mars. Having chilled chamber walls makes it difficult to use a carbon dioxide atmosphere, because the CO_2 will freeze out on the cold surfaces (an effect that plagued tests on Mars Pathfinder[12]), and so the chamber was back-filled with 10 mbar of nitrogen

Fig 5.9   Helicopter mounted on a stand in the 25-ft space simulation chamber. The blue-black material covering the box is a metallized Kapton designed to maximize the absorption of sunlight and reduce the loss of radiant heat to keep the system warm on Mars.
Source: JPL/Caltech.

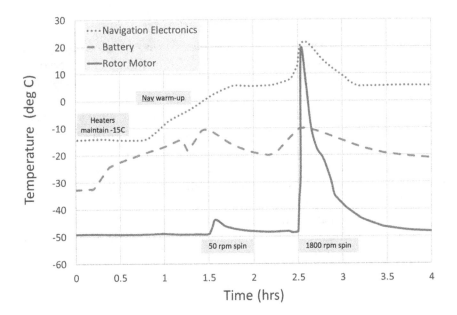

Fig. 5.10 Helicopter temperature evolution during thermal test in the 25-ft chamber at Mars pressure (N_2). There is appreciable warmup of the navigation electronics during operation, whereas the much more massive battery responds more slowly to power dissipation. The propulsion motor warms up dramatically during rotor operation.
Source: Author, from data in [13].

instead. The convective heat transfer, in any case, cannot be perfectly simulated in that buoyant (free) convection is stronger in Earth's gravity than in Mars's when wind is not blowing, and is dominated by the wind when it is. Part of the thermal testing was done in vacuum to be able to isolate the gas-mediated heat transfers. (Gas conduction is important internal to the vehicle, as well.)

Another testing and analysis challenge was that the helicopter itself has only a handful of temperature sensors for housekeeping monitoring on Mars, and the data from these are only available when the electronics are powered up. Thus (as in most spacecraft thermal balance tests) additional thermocouples are installed to continuously monitor temperatures (Fig. 5.10) all over the vehicle throughout the test, whether the vehicle is powered or not. But the cabling to these additional sensors introduces artificial conductive heat leak paths that are not present on the vehicle on Mars,‡ and so the mathematical model that is tuned to the results of the test must be corrected for these effects.

The flight model (FM) finished acceptance testing in early 2019.

---

‡One is reminded of Schrodinger's uncertainty principle—you cannot measure a system without perturbing it!

COMPONENT TESTING

Before the system-level tests of the vehicle as a whole, various components had to have their own tests; for example, the gearbox design of the servos was tested for over 1500 h of continuous operation under load to show that it would last without failure or significant degradation.

Being a technology demonstration mission without the same exhaustive reliability class as Perseverance gave Ingenuity the ability to use commercial (rather than previously space-proven) parts; however, such components could not be used without some qualification. Thus, environmental tests and up-screening (buying large lots of parts and testing them, choosing only the best among them) were employed.

There is also a motto—"never trust the data sheet"—that motivates doing independent performance tests if the budget and schedule allow. For example, the miniature BMI-160 inertial measurement unit (IMU) angular rate sensor had a specification for angular random-walk error of 0.007 rad/s/root Hz, which would be sufficient to meet the requirements of Ingenuity's short flight; however, early IMU testing revealed that the sensor was sensitive to vibration. The sensor's susceptibility to accelerations causing a gyro bias would produce an unacceptable 9-deg error over a 90-s flight.§ This meant that the state vector modeled in the navigation algorithm [minimal augmented state algorithm for vision-based navigation (MAVeN)] had to be extended from 12 to 21 dimensions (adding three for current attitude, three for base attitude, and three for gyro bias). To practicably implement this correction, however, relies on the assumption of flat terrain, which of course is not perfectly true for real planetary surfaces.[13]

Although in principle the performance of antennas can be computer-modeled, the interaction of the radio-frequency fields with the ground and reflections from nearby structures can be difficult to fully capture. Thus, field tests of a rover mockup were used to confirm the computer estimates. From some directions the antenna gain was low enough[14] that it would have been preferred to add a second antenna to fill the gaps, but this could not be accommodated. Knowing the performance accurately, however, would allow the rover operators to avoid particularly unfavorable orientations.

PLANETARY PROTECTION

There are international agreements on various aspects of spaceflight, notably the 1967 Outer Space Treaty. Among the stipulations is that space exploration should avoid the harmful contamination of celestial bodies ("forward

---

§T. Tzanetos indicated to me that in fact the BMI160 gyro issue was a resonance issue, with the signal spiking as the rotor blades ramp up to their operational rpm, and it only occurs during spin up and spin down, not during flight, so perhaps was not as serious as the peak error rate might imply. It only occurs at Mars temperatures, highlighting the need to test components at flight-relevant conditions.

contamination") and also any adverse effects to Earth's environment resulting from any return of extraterrestrial material. To ensure compliance with these agreements, NASA has an official called the Planetary Protection Officer who develops standards and processes, tailored for each planetary body and mission, for the analysis of these issues.

For Mars, where there is at least a reasonable possibility of extant life, or at least traces of previous life, the planetary protection standards are rather exacting. The *bioburden*, or number of bacterial and other spores on spacecraft surfaces, must be evaluated by standardized protocols using swabs and wipes of the hardware, and then culturing the samples to see how many spores grow. Even though spacecraft are assembled in clean rooms with filtered air handling, there may a substantial number of bacteria, fungi, and so on. As Keennon[3] put it "They came out with their globes and petri dishes and swabs, inspecting and taking samples from our work areas. It was intense." Components or assemblies are often cleaned with alcohol, or baked in an oven to reduce the bioburden, as well as to remove any molecular contamination such as skin oils or adhesive residues (Fig. 5.23).

At launch, the systems going to Mars (Perseverance, Ingenuity, and the cruise stage, aeroshell, and descent stage) were allocated to carry fewer than 500,000 bacterial spores, of which no more than 300,000 spores were permitted on the elements reaching the Martian surface. The helicopter was allotted 33,000 spores and ended up with a count of 21,900. (These values are estimated by scaling up from the sampled areas—each bacterium isn't actually counted!)

DELIVERY AND INTEGRATION

While the helicopter and rover were being built at AeroVironment and the Jet Propulsion Laboratory (Fig. 5.11), both in Southern California, some other elements were being readied in other parts of the United States. One key part was the multimission radioisotope thermoelectric generator (MMRTG) (Fig. 5.12), which would provide electrical power and heat to the Perseverance rover. The precious plutonium dioxide fuel was loaded into this unit at the Idaho National Laboratories (INL) in winter 2019/2020, and the generator underwent tests for several weeks before being transported by road to Florida.

The formal addition of the helicopter to the Perseverance rover took place rather late in the rover's development. The design and provision of the deployment mechanism, protective cover, and mounting structure were contracted to Lockheed Martin in Denver, Colorado, which also built the aeroshell for Mars 2020.

Because this system was added late in the rover's development, there were relatively few available unallocated command channels, and so the deployment mechanism (Fig. 5.13) had to be artfully designed to operate with just a few

Fig. 5.11 Assembly operations are rehearsed in meticulous detail to train personnel. Here, the helicopter and mounting plate are mated to the inverted underside of the rover. Onlookers are just visible in the viewing gallery at the top edge of the image.
Source: NASA/JPL/Caltech.

distinct signals to actuate the separation bolt, pyrotechnic cable cutter, and deployment motor.

In spring 2020, just as the COVID-19 pandemic began, the final integration of the Mars 2020 mission began at Kennedy Space Center in Florida. The

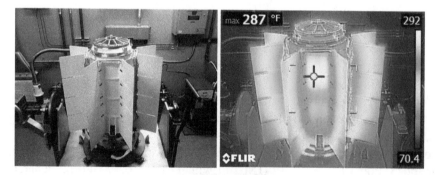

Fig. 5.12 The Perseverance MMRTG at the Idaho National Labs in March 2020.¶ The plutonium fuel produces about 2 kW of heat, hence the warm surfaces. This heat is essential for vehicles in cold environments such as Mars or Titan.
Source: INL; photos acquired at the direction of the author.

¶In fact, a visit to INL to make Dragonfly-motivated measurements on the Perseverance MMRTG was my last trip just before business travel was suspended by the COVID-19 pandemic. I was very fortunate to be permitted to make measurements on the atmospheric ionization in the vicinity of the generator in the fleeting opportunity between when it was fueled and when it would be shipped to Cape Canaveral. I was close enough to feel the generator's Promethean warmth on my cheek. The 12-millirem radiation dose I acquired during my day in the facility was little more than the 7 or so millirem that I acquired flying to and from Idaho from the East Coast.

Fig. 5.13 **The Mars helicopter on its deployment mechanism and its mounting structure (see also Figs. 3.2 and 3.6) ready for a vibration test. An anodized aluminum structure restrains two of the helicopter legs at top. Notice the (nonflight) blue wires attached with Kapton tape: at the end of each wire is a small block, which is an accelerometer sensor to measure the response of the structure to vibration or pyrotechnic shock. After the tests, these sensors and wiring are removed.**
Source: Lockheed Martin.

pandemic caused considerable disruption to NASA programs (as well as everything else!), and only mission-critical agency travel was permitted. The integration activities (Fig. 5.14), of course, fell under this category.

Ingenuity, the Perseverance rover, the rocket-powered descent stage that would deliver them to the surface, the aeroshell that would protect these elements during hypersonic entry, and the cruise stage that would support and guide the aeroshell to Mars were all brought together (Fig. 5.15) at Cape Canaveral.** As well as making and testing all the mechanical and electrical connections between these elements, various hazardous items would be installed, such as pyrotechnic devices for separation events, the hydrazine rocket propellant, and (last of all) the MMRTG that would power the rover.

The helicopter and its delivery system were attached in a day-long operation on 6 April 2020, after the descent stage had been fueled. There were 34 electrical connections between the helicopter and its delivery system on the rover, and these were checked to establish that the rover could charge the helicopter battery and send and receive signals to the helicopter.

**NASA's Kennedy Space Center is formally only part of the facility, much of which is Cape Canaveral Air Force Station.

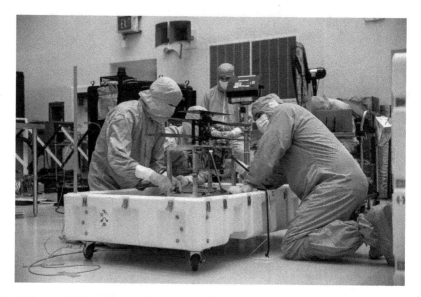

Fig. 5.14 Ingenuity is inspected on its rigid mounting on its shipping container. Note the clean room suits as well as the grounding cables at lower right to avoid any electrostatic discharge that could damage the helicopter's electronics. Humidifiers and air ionizers are also used to help allow any electrical charges to bleed away.
Source: NASA.

Fig. 5.15 Ingenuity on its yellow mounting plate, mated on 6 April 2020 to the belly of the Perseverance rover at Kennedy Space Center. The rover wheels are in a retracted position to fit in the aeroshell and are wrapped in metallized film covers for cleanliness. These are removed just before final encapsulation in the aeroshell. At right, note the gold-colored *U*-shaped pipes; these "catch" radiant heat from the MMRTG to allow this warmth to be drawn into the vehicle, using a pumped fluid.
Source: NASA/JPL/Caltech.

Fig. 5.16 Inside the Payload Hazardous Servicing Facility at NASA's Kennedy Space Center in early May 2020. The eight red cylinders are remove-before-flight covers on the descent stage thruster nozzles, and the lozenge-shaped silver item at the bottom with dark circles is the descent stage Doppler radar. The helicopter is encapsulated in its black "violin case" cover at upper right, and the rover wheels still have covers.
Source: NASA/Christian Mangano.

The rover, in turn, was integrated with the descent stage and the backshell (Fig. 5.16), and then that assembly was mated with the cruise stage and the front heat shield of the aeroshell (Fig. 5.24). This assembly was then attached (upside down, in effect) to the launch vehicle adapter and the payload fairing mated around it (Fig. 5.17). The fairing was lifted by a gantry to mount on top

Fig. 5.17 The Mars 2020 aeroshell and cruise stage, about to be encapsulated in the cavernous payload fairing for the Atlas V launcher on 18 June 2020.
Source: NASA/KSC.

Fig. 5.18 **The Atlas V rocket carrying Perseverance and Ingenuity blasts off from Space Launch Complex-41 at 0750 EDT on 30 July 2020. Although this happened to be a beautiful clear day, the three towers around the pad are designed to preferentially attract any lightning to avoid direct strikes onto the launch vehicle.**
Source: NASA.

of the Atlas V 541 launch vehicle stack.[††] Finally, within a few days of liftoff, the MMRTG was installed through special hatches in the aeroshell and payload fairing.

LAUNCH AND CRUISE

The launch window, when appropriate planetary alignments would permit flight to Mars, opened on 17 July 2020 and lasted through 15 August 2020. Due to some technical delays at the Cape, Mars 2020 launched on 30 July 2020 (see Fig. 5.18).

The 293-million-mile (471-million-km) cruise was uneventful. Every two weeks controllers checked the state of charge of Ingenuity's batteries, typically topping them off to the 35% considered to be optimal for preserving the cells' capacity.[15]

The mission arrived at Mars on 18 February 2021. The thin Mars atmosphere forces the arrival events to be compressed into "seven minutes of terror," beginning with the hypersonic entry in the aeroshell (see Fig. 5.19). Although in many missions this is a passive (ballistic) phase, with drag being

---

[††]The 541 designation implies that the vehicle has a 5-m-diameter payload fairing, that four strap-on boosters augment the thrust at liftoff, and that a single Centaur upper stage engine is used.

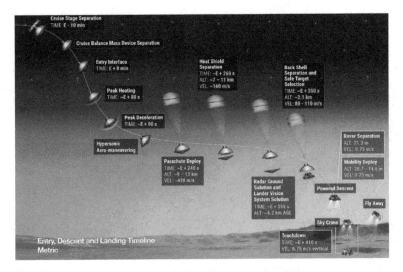

Fig. 5.19   Key events during the entry, descent, and landing.
Source: JPL.

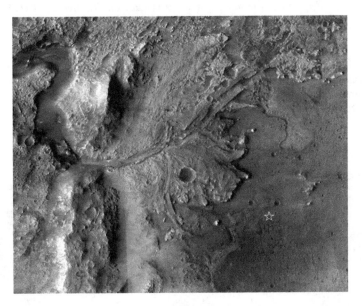

Fig. 5.20   An image composite of Jezero Crater on Mars, the landing site for NASA's Mars 2020 mission. The color is false, based on near-infrared observations by the compact reconnaissance imaging spectrometer for Mars (CRISM), which highlight different minerals such as clays and carbonates. These minerals and the morphology of the river delta deposit suggest the crater was filled with a lake for some time (and thus may have been a habitable environment). The star shows the Perseverance landing site, named after the science fiction author Octavia E. Butler. A tongue of dark brown material at 7 o'clock to the landing site is the ripple field named Séítah.
Source: NASA/JPL-Caltech/MSSS/JHU-APL (Malin Space Science Systems/Johns Hopkins University Applied Physics Laboratory).

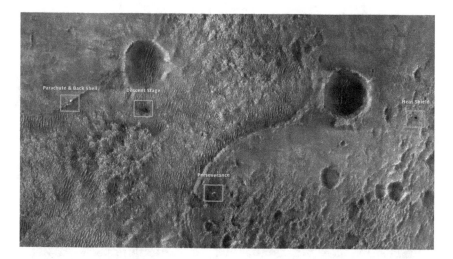

Fig. 5.21   This first image of NASA's Perseverance Rover on the surface of Mars from the high resolution imaging experiment (HiRISE) camera aboard NASA's Mars reconnaissance orbiter (MRO) shows the many parts of the Mars 2020 mission landing system that got the rover safely on the ground. The image was taken on 19 February 2021. This annotated version of the image points out the locations of the parachute and backshell, the descent stage, the Perseverance rover, and the heat shield. Each inset shows an area about 650 ft (200 m) across. The rover sits at the center of a blast pattern created by the hovering descent stage that lowered it there using the sky crane maneuver. The descent stage flew off to crash at a safe distance, creating a V-shaped debris pattern that points back toward the rover. Earlier in the landing sequence, Perseverance jettisoned its heat shield and parachute, which can be seen on the surface in the separate locations illustrated.
Source: NASA/JPL-Caltech/University of Arizona.

Fig. 5.22   A later Mars-pressure flight test using an EDM (this time with solar panel). Note the tether coiled to the right.
Source: NASA.

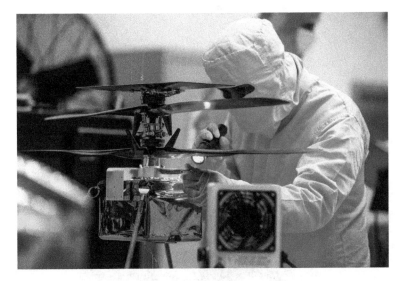

Fig. 5.23   Space hardware must be kept scrupulously clean. Close inspection with good lighting is a simple but effective means of detecting particulate debris. As is familiar from crime scene investigation TV shows, ultraviolet light can also indicate organic contamination via fluorescence.
Source: NASA/JPL/Caltech.

the only significant force, Mars 2020 developed appreciable lift and used thrusters to angle the aeroshell to bleed off energy at the optimum rate to reach the desired target area. By flying in this way, the uncertainty or *delivery ellipse* of Mars 2020 was very small, only 7.7×6.6 km,[‡‡] allowing plans for the landed mission to be closely focused on the target area, the Jezero impact crater and the delta deposits at its western margin (see Fig. 5.20). As with Curiosity before it, a large parachute was deployed by a mortar at about 10-km altitude and a speed of Mach 1.5 to stabilize the aeroshell before it slowed to transonic speed. The large chute allowed the heat shield to safely separate and quickly bled off speed to about 100 m/s about 2 km above the ground. The descent stage carrying the rover then dropped away from the backshell and parachute, and flew under rocket power down to the surface. One development beyond Curiosity was that Mars 2020 used terrain-relative navigation. A map of hazards (large rocks and sand dunes that might trap the rover by allowing wheel slippage) had been constructed based on orbital imaging, and a camera system would identify where the spacecraft was in this map and fly (using its rocket thrusters) to a known safe area (Fig. 5.21).

[‡‡]Actually, various values are quoted, depending on the stage of the project at which they were estimated. Regardless of which benchmark (at a development review, at launch, or just prior to entry) is used, Mars 2020's delivery ellipse was much smaller than any previous mission's.

Fig. 5.24 **The Mars 2020 cylindrical cruise stage sits atop the white conical back shell. A spherical gold rocket propellant tank is visible top center, and the white curved panels are radiators to reject the heat pumped from the MMRTG. The backshell contains the powered descent stage and Perseverance rover (now with uncovered wheels). The helicopter in its cover is visible on the lander belly. Below it is the tan-colored heat shield that is about to be attached to the back shell. The image was taken on 28 May 2020.**
Source: NASA/JPL-Caltech/Kennedy Space Center (KSC).

The landing was successful, and the entry, descent, and landing (EDL) was better-documented than previous missions, with dramatic movie footage acquired by upward-looking cameras to observe the parachute deployment and inflation, and downward-looking cameras to show the deployment of the rover on a tether beneath the skycrane and the interaction of the thruster plumes with the dust and sand on the surface. Although the data volume associated with the hundreds of image frames of these movies was formidable (hundreds of gigabytes), the buildup of orbital relay assets at Mars allowed these data to be sent within a few days. Once the rover systems were checked out, the descent software was replaced with that for landed operations, and the initial scientific assessment of the terrain around the landing site was made. The stage was then set for helicopter operations.

REFERENCES

1   Balaram, J., and Tokumaru, P. T., "Rotorcrafts for Mars Exploration," *11th International Planetary Probe Workshop*, 2014, https://www.hou.usra.edu/meetings/ippw2014/pdf/8087. pdf

2 Cooper, A., "Perseverance Rover, Ingenuity Helicopter, and the Search for Ancient Life on Mars," *60 Minutes*, 9 May 2021, https://www.youtube.com/watch?v=gRrFRL5v0ig

3 Peck, A. "Inside Ingenuity with AeroVironment," 29 June 2021, https://insideunmanned-systems.com/inside-ingenuity-with-aerovironment/

4 Many fascinating interactions of aerodynamics and gyrodynamics are discussed in my book *Spinning Flight: Dynamics of Frisbees, Boomerangs, Samaras and Skipping Stones*, Springer, 2006.

5 Grip, H. F., Johnson, W., Malpica, C., Scharf, D. P., Mandić, M., Young, L., Allan, B., Mettler, B., Martin, M. S., and Lam, J., "Modeling and Identification of Hover Flight Dynamics for NASA's Mars Helicopter," *Journal of Guidance, Control, and Dynamics*, Vol. 43, No. 2, 2020, pp. 179–194. See also Grip, H. F., Scharf, D. P., Malpica, C., Johnson, W., Mandic, M., Singh, G., and Young, L. A., "Guidance and Control for a Mars Helicopter," *2018 AIAA Guidance, Navigation, and Control Conference*, p. 1849.

6 Smeaton, J., "An Experimental Enquiry Concerning the Natural Powers of Water and Wind to Turn Mills, and Other Machines, Depending on a Circular Motion," *Philosophical Transactions of the Royal Society*, Vol. 51, 1759, pp. 100–174. doi:10.1098/rstl.1759.0019.

7 Caltech, "How Do You Test a Helicopter Bound for Mars?," 22 April 2021, https://www. caltech.edu/about/news/how-do-you-test-a-helicopter-bound-for-mars

8 Veismann, M., Dougherty, C., Rabinovitch, J., Quon, A., and Gharib, M., "Low-Density Multi-fan Wind Tunnel Design and Testing for the Ingenuity Mars Helicopter," *Experiments in Fluids*, Vol. 62, No. 193, 2021, https://link.springer.com/article/10.1007/s00348-021-03278-5

9 Escobar, D., "Fundamental Understanding of Helicopter Aeromechanics on Mars Through Chamber Testing and High-Fidelity Analysis," Ph.D. dissertation, Oct. 2000, https://doi.org/10.13016/r0s8-ltmw; see also Escobar, D., Chopra, I., and Datta, A., "High-Fidelity Aeromechanical Analysis of Coaxial Mars Helicopter," *Journal of Aircraft*, Vol. 58, No. 3, 2021, pp. 609–623.

10 Pipenberg, B. T., Keennon, M., Tyler, J., Hibbs, B., Langberg, S., Balaram, J., Grip, H. F., and Pempejian, J., "Design and Fabrication of the Mars Helicopter Rotor, Airframe, and Landing Gear Systems," *AIAA Scitech 2019 Forum*, p. 620.

11 Cappucci, S., and Pauken, M. T., "Thermal System and Environmental Testing of the Mars Helicopter," *International Conference on Environmental Systems*, ICES-2020-95, 2020. See also Schmidt, T. M., Cappucci, S., Miller, J. R., Wagner, M. F., Bhandari, P., and Pauken, M. T., "Thermal Design of a Mars Helicopter Technology Demonstration Concept," *48th International Conference on Environmental Systems*, ICES-2018-018, Albuquerque, New Mexico, 8–12 July 2018.

12 Lorenz, R. D., "Atmospheric Test Environments for Planetary In-Situ Missions: Never Quite 'Test as You Fly,'" *Advances in Space Research*, Vol. 62, No. 7, 2018, pp. 1884–1894.

13 Bayard, D. S., Conway, D. T., Brockers, R., Delaune, J. H., Matthies, L. H., Grip, H. F., Merewether, G. B., Brown, T. L., and San Martin, A. M., "Vision-Based Navigation for the NASA Mars Helicopter," *AIAA Scitech 2019 Forum*, 2019, p. 1411.

14 Chahat, N., Miller, J., Decrossas, E., McNally, L., Chase, M., Jin, C., and Duncan, C., "The Mars Helicopter Telecommunication Link: Antennas, Propagation, and Link Analysis," *IEEE Antennas and Propagation Magazine*, Vol. 62, No. 6, 2020, pp. 12–22.

15 NASA, *Ingenuity Mars Helicopter*, Landing Press Kit, January 2021.

Chapter 6

INGENUITY OPERATIONS

"Far better it is to dare mighty things . . . than to rank with those poor
spirits who neither enjoy nor suffer much because they live in the gray
twilight that knows neither victory nor defeat."

—President Theodore Roosevelt (1858–1919)*

In this chapter, we review what the Ingenuity helicopter actually has done
on Mars.[1] According to premission plans,[2] the technology demonstration
objectives of Ingenuity would be met with five flights of progressively higher
complexity, performance, and risk (see Table 6.1 and Fig. 6.1). The time of
day of flights was influenced by helicopter factors such as wanting the battery
and other temperatures to be at reasonable values and the battery to have an
adequate state of charge. The timing of the postflight data relay [ie, when the
on-board telemetry would be sent to the base station on the rover, when it
would be sent from the rover to an orbiting satellite such as the European
Space Agency's (ESA's) Mars Trace Gas Orbiter or NASA's Mars
Reconnaissance Orbiter, and when it would be sent from there to Earth] also
had to be taken into account in planning. Also coming into play were expectations (initially from meteorological models simulating the weather in
Jezero crater, and later direct measurements by instruments on the
Perseverance rover) of the winds and atmospheric density/temperature. One
tradeoff considered as data emerged was that the time of day with weaker
background winds might actually be more gusty. Note, however, that there was no
real-time weather evaluation—in effect, Ingenuity was commanded to attempt to
perform a flight at a given time on a given Mars day (sol), whatever the weather
would happen to be. Although the rover might take weather measurements that

---

*The "Dare Mighty Things" motto often cited by the Jet Propulsion Laboratory (JPL) was encoded in
the parachute markings of the Perseverance rover—see Fig. 3.10 in Chapter 3.

TABLE 6.1 TECHNOLOGY DEMONSTRATION FLIGHT GOALS

Flights and Flight Description			Test Conditions
1	Baseline Case	Ascend, hover, and land. • Up/down to ~3 m AGL (Above Ground Level)	Least windy time of day (eg, 1100) Same terrain in view during entire flight
2	Baseline Case	Fly higher and fly ~5 m laterally. • ~5 m AGL • Slow lateral motions	
3	Baseline Case	Fly 50 m out and back. Take images of potential new landing sites.	New terrain in view as flight progresses Potentially fly in moderate wind conditions
4 5	Stretch Goals*	Fly to new landing site. Fly away 500 m. Fly in higher winds. Fly in early morning or late evening. Fly over rough terrain. Fly at higher altitudes.	New terrain in view as flight progresses Potentially fly in moderate wind conditions Late afternoon eddy wind formations Low sun angle lighting Challenging slopes and rocks

*The stretch goals were representative challenges, but were not individually required for mission success.

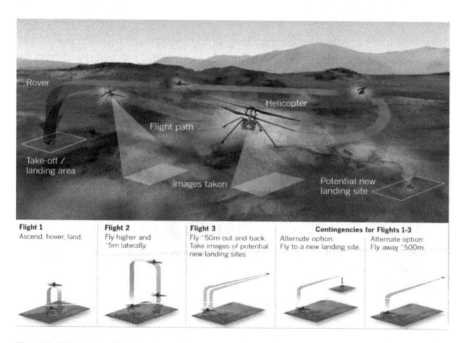

Fig. 6.1 Schematic of technology demonstration flights.
Source: NASA.

would inform what the flight conditions were after the fact, there was no process to abort a takeoff in real time due to weather variations.

DELIVERY

Before the technology flights could be executed, Perseverance first had to identify a suitable area (airfield) (Fig. 6.2) to conduct the Ingenuity technology demonstration mission. Documenting the site with cameras and then providing the communication support for the flights would require Perseverance to remain nearby. The sooner the technology demonstration could be completed, the sooner Perseverance could move off on the much more serious business of its primary mission of sample acquisition for eventual return to Earth, a mission that would require tens of kilometers of driving.

Engineers looking for a desirable airfield considered the following criteria:

- *Helipad:* A 3-m box where Perseverance would deploy the Ingenuity Mars helicopter. Preferably, the box should be free of rocks more than 5 cm (2 inches) high and should be sloped no more than an average of around 5 deg.
- *Airfield:* A 10-m (33 ft) box surrounding the helipad. The surface should have as few rocks as possible exceeding 5 cm in height and have an average slope of no more than approximately 4 deg.
- *Flight zone:* This oval area is defined by a boundary that extends horizontally 15 m in any direction beyond the Mars helicopter's expected flight path during test flights. The terrain inside the zone should have sufficient visual features for the helicopter to track its movements using the on-board navigation camera.

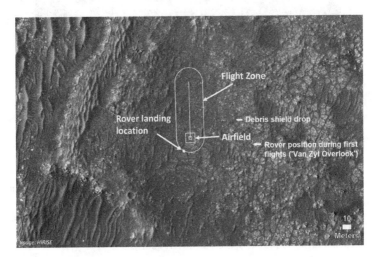

Fig. 6.2 Helicopter flight zone, marked on a satellite image of Jezero. Note the sand ripples of Seitah to the right of the image.
Source: University of Arizona/NASA/JPL-Caltech.

The planning also required the identification of an observation location for the Perseverance rover, preferably relatively level and about 100 m from the edge of the flight zone to ensure there was no danger of Ingenuity striking the rover and threatening its subsequent mission. The observation location[†] should have direct line-of-sight to the entire flight path, both to ensure good radio communication and to permit the rover's high-resolution mast-mounted zoom camera (Mastcam-Z, aka ZCam) to observe the flights.

Examining data from orbital images (now knowing, postlanding, where the rover actually was!) and from rover images taken during descent and once on the ground, a suitable site was quickly identified. But before deploying the helicopter onto the ground, the debris shield that protected Ingenuity from debris kicked up by the skycrane rocket exhaust[‡] had to be jettisoned (see Fig. 6.3).

Perseverance dropped the debris shield protecting Ingenuity on 21 March 2021. Several intermediate steps in the helicopter deployment were then executed,[3] each on a subsequent sol (see Fig. 6.4). First, a Frangibolt that had secured Ingenuity tight against the rover body was fired. This is a bolt with a heater designed to cause stress via thermal expansion, which breaks a brittle element. (This type of release has some advantages over pyrotechnic bolts in cases where the exact time of actuation on timescales of seconds-minutes is not important.) On the next sol, a pyrotechnic guillotine cut a cable, releasing

Fig. 6.3 The debris shield, a protective covering on the bottom of the Perseverance rover, was released on 21 March 2021, or sol 30, of the mission. This image was taken by the wide angle topographic sensor for operations and engineering (WATSON) camera on the scanning habitable environments with Raman and luminescence for organics and chemicals (SHERLOC) instrument, located at the end of the rover's long robotic arm.
Source: NASA/JPL/MSSS (Malin Space Science Systems).

[†]This location was nicknamed "twitcher's point" after a British term for a birdwatcher.
[‡]One of the wind sensors on Curiosity was broken at landing; this was presumed to be due to a small rock thrown up by the exhaust plume impinging on the ground.

Fig. 6.4 Ingenuity being deployed from the belly of the Perseverance rover on sol 39, 30 March 2020. These images were taken by the WATSON camera.
Source: NASA/JPL-Caltech/MSSS.

an arm which allowed Ingenuity's weight to swing it down about 45 deg away from the rover. This movement allowed two of the legs to deploy using springs. Then an actuation motor latched Ingenuity into the vertical position. where it was still held by a single bolt onto the rover, its feet a mere 13 cm above its destination planet. The helicopter was still directly connected electrically to the rover. On the fourth deployment sol, the last two legs were released, and the deployment configuration was verified by a camera mounted on the rover's arm, which was able to look under the rover's belly (see Fig. 6.5).

Fig. 6.5 This low-resolution view of the floor of Mars' Jezero crater and a portion of two wheels of NASA's Perseverance Mars rover was captured by the RTE (Return to Earth) color imager aboard Ingenuity. The image was taken on 3 April 2021, while the solar-powered rotorcraft was still beneath the rover after being deployed.
Source: NASA/JPL-Caltech.

The last release operations were to top off the battery and conduct a final check before release (including taking a picture with its cameras, while still attached). It would be important to back the rover away quickly once the helicopter was on the ground and electrically on its own, because its solar panel would not develop any power while in the rover's shadow, and its battery would not last more than one night.

Ingenuity was deployed on April 3. The rover then backed away 5 m and inspected Ingenuity with its cameras (see Fig. 6.6). Although the helicopter had apparently deployed nominally, it appeared that there was dust on its solar panel (see Fig. 6.7), which presumably had managed to be blown around the edges of the debris shield at landing. (It was not sealed, but rather was to prevent the impact of pebbles launched by the impingement of the retrorocket blast on the ground. Such a pebble impact was suspected to have caused the failure of one of the Curiosity rover's wind sensors at landing.)

Ingenuity's rotor blades were successfully unlocked on 8 April 2021 (mission sol 48), an event observed by Perseverance's cameras (the unlocking action required both rotors to be rotated in the same direction), and the helicopter later performed a low-speed rotor spin test at 50 rpm. On April

Fig. 6.6 Perseverance took a selfie with the Ingenuity helicopter, seen here about 3.9 m (13 ft) from the rover in this image taken 6 April 2021 (sol 46) by WATSON, located at the end of the rover's long robotic arm. Perseverance's selfie with Ingenuity is made up of 62 individual images stitched together once they were sent back to Earth; Developing the complex sequence of commands to position the camera to achieve the image overlaps subject to the constraints on the arm movements, and to verify this sequence on a testbed rover on Earth, took days and person-weeks of effort.
Source: NASA/JPL-Caltech/MSSS.

Fig. 6.7 A sequence of Mastcam-Z images on 8 April 2021 (sol 48) showed the blades performing a wiggle test to demonstrate movement before the actual spin-up. Something of a surprise was the amount of dust on the solar panel.
Source: JPL/NASA/Caltech/Cornell.

16, Ingenuity successfully passed the full-speed 2537-rpm rotor spin test while remaining on the surface. It would make its first flight three days later.

HOW INGENUITY FLIES

The propagation delays of communication mean that each flight must be a predefined sequence followed by Ingenuity's on-board control system.[4] The flight is managed by an on-board process called the Mode Commander, which is implemented as a state machine, with the transition from one mode to another occurring only when certain conditions are met.

The modes are as follows[5]:

- *Idle:* Do nothing, wait for commands from ground station.
- *Inclinometer averaging:* Gather and average inclinometer data to pass to the navigation subsystem
- *Initialize estimator:* Wait for navigation to finish initialization, as indicated by low estimation errors.
- *Spinup:* Rotor spinup to setpoint speed.
- *Takeoff preparation:* Wait for Navigation to start full 6-degree-of-freedom inertial measurement unit (IMU) integration, temporary open-loop application of nonzero thrust below takeoff thrust, to reduce magnitude of motor load increase on takeoff.

- *Takeoff:* Climb to 5-cm altitude or until timeout reached, with attitude rate-only control to avoid any adverse interactions between the control system and the ground.
- *Climb:* Climb to flight altitude with full control.
- *Waypoint tracking:* Translate between waypoints. (Hover is simply remain at a waypoint, with altitude hold.)
- *Descent:* Descend while maintaining horizontal position (Visual feature tracking stops at 1m altitude).
- *Landing:* Continue descent with touchdown detection armed.
- *Touchdown:* Disable control to avoid ground interactions; reduce thrust by setting collectives to lowest setting.
- *Spindown:* Spin down rotors.

Flights can be aborted if a fault is detected or by command from the ground station. (In practice, this latter mode was only meaningful in testing on Earth since obviously the two-way light time is much longer than any flight.) Among the on-board fault detections was a watchdog timer function in the field programmable gate array (FPGA) that checked for a signal expected at regular intervals from the flight controller microcontrollers; if it was not present, it declared a fault.[6] This watchdog timer expiration error occurred on sol 49 (April 9) during a planned high-speed rotor spin test, preventing the state machine from transitioning into the flight state. A software fix was identified within a few days, but because the timing glitch that triggered the error was expected only 15% of the time, the project proceeded at risk, rather than wait for the software update to be uploaded to Mars.

In order to implement these various modes, the autopilot in the flight controller executes a guidance and control loop at a rate of 500 times/s, based on IMU measurements at that rate. Navigation information comes in 30 times/s from the navigation computer, implemented with a 2.24-GHz Snapdragon processor. The FPGA handles the communication between the sensors and the active flight controller, and switches to the backup flight controller if it detects a fault. The flight controller develops actuator commands to guide the vehicle according to the dynamics expected based on the system identification experiments on Earth. In doing this stably, the various millisecond delays in sensor response and communications latency must be taken into account, as must the finite response time of the servo actuators for the swashplate, which are modeled as second-order systems with a bandwidth of 12 Hz and 85% damping.

This flight control implementation was intended to provide robust performance, even given some uncertainty in the models, and to achieve sufficiently accurate station-keeping performance in response to horizontal gusts (maximum displacement of 2 m in response to a 5 m/s gust).

FIRST FLIGHT: THE WRIGHT BROTHERS' MOMENT ON MARS

After the helicopter was deployed, the rover retreated to a safe distance (64 m), the "twitcher's point" being designated Van Zyl overlook.[§] Although the first flight had been intended for April 11, the watchdog timer bug mentioned earlier had aborted the rotor spin test on sol 49, which had prevented the helicopter from entering the flight state. Once the issue was diagnosed and determined to have only a small chance of recurrence, the flight was replanned for April 19 at 0734 GMT, or 1233 Local Mean Solar Time.[¶] The vehicle spun up its rotors, climbed to 3-m altitude, hovered for 5 s in a fixed orientation, then executed a 96-deg yaw and hovered for another 20 s before touching down again after 39.1 s.[**]

The on-board sensor data (Fig. 6.9) were transmitted from the helicopter to the rover and from there beamed back to Earth, where they were received about 3 h after the flight. What most people saw, however, was the documentation of the flight by the zoom camera Mastcam-Z on the rover (see Fig. 6.8), which was able to acquire a remarkable movie of the flight. In fact, Ingenuity had a tangible link to its terrestrial counterpart—attached to its body was a tiny piece of muslin cloth used in the wing of the Wright Flyer.[††]

The color RTE camera was not exercised on the first flight; however, black-and-white navigation nadir-pointed navigation (NAV) camera images, used to

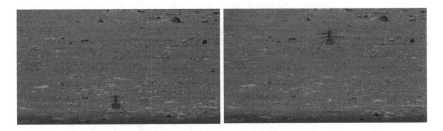

Fig. 6.8 The Wright brothers' moment on Mars: two frames of a movie sequence of the first flight of Ingenuity, acquired on 22 April 2021 by the left Mastcam-Z camera on the rover mast.
Source: NASA/JPL-Caltech/Arizona State University/MSSS.

[§]This was named after Jacob Van Zyl, a prominent JPL radar scientist who passed away unexpectedly in August 2020.

[¶]There are a couple of time systems in common use, necessitated by the orbital eccentricity of Mars, which makes the seasons of appreciably unequal length and causes a solar day (the interval between solar meridian crossings) to be longer near southern summer. Local Mean Solar Time (LMST) differs from Local True Solar Time (LTST) in that it is prescribed to have days of uniform length, which is more convenient for operations planning.

[**]For reference, the Wright brothers' first powered flight in 1903 was only 12 s long and reached only 2.4 m.

[††]In fact, in 2017 I had, independently, had a similar idea that such a piece of fabric might make a useful calibration target for Dragonfly's microscopic imager. A number of pieces of Wright Flyer muslin have also been to the moon, carried in the personal kit of Neil Armstrong; these are now in the hands of private collectors.

develop real-time horizontal velocity information for the autopilot, were sent back. Some interesting features were noted in the NAV camera images, which because the flight was conducted around noon when the sun was nearly verti-cally above Ingenuity, and the camera looks directly downwards, showed Ingenuity's shadow (Fig. 6.10).

The relatively short exposure time meant that the shadow of the rotor blades was not blurred appreciably. Yet there was the odd feature that the shadow of the rotors was less deep (dark) than the shadow of the helicopter body. This was due to the architecture of the camera detector, which has no physical shutter. The lines of the image are moved to a storage area on the detector that is in principle masked off so no light gets through, but in prac-tice there is some light leak. Thus, the brightness of each pixel is a weighted sum of the light that came into an operating (unmasked) detector pixel dur-ing the planned exposure, plus whatever light leaked through during the longer storage period on a masked pixel. Now, during the storage period, the shadow of the fast-moving blade had moved on, and so light leaked in and

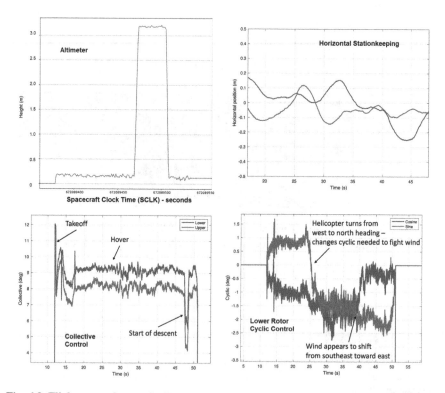

Fig. 6.9 Flight control records, like the data from a black box, from Flight #1. The on-board altitude estimate shows an offset and some measurement noise. The vehicle's cyclic control maintained its horizontal movements closely around the desired position.
Source: Author compilation of NASA/JPL/Caltech images.

Fig. 6.10 The downward-looking navigation camera on Ingenuity took this shot while hovering over the Martian surface on 19 April 2021, during the first instance of powered, controlled flight on another planet. Bright rock slabs and fragments are visible against the darker regolith. The tread marks of Perseverance are visible, darker still, below and above the helicopter shadow. Notice that the shadows of the helicopter solar panel and legs are darker than the shadows of the rotor blades.
Source: NASA/JPL/Caltech.

diluted the blade shadow. However, because the helicopter as a whole had barely moved during the storage interval, the shadow of the helicopter remained in the same place on the detector, and so that shadow was not diluted by light leak.

TECHNOLOGY DEMONSTRATION FLIGHTS

In subsequent flights, the distances and speeds were progressively increased to meet the technology objectives in the plan. The second flight performed a horizontal translation as well as vertical ascent and descent, and also operated the color RTE camera (see Figs. 6.11 and 6.12).

Even though the NAV images were small (one-third megapixel) and were all stored on board the helicopter, because there were some 30 frames taken per second, only a small subset would be generally transmitted to the rover over the 20-kbps radio link and then to Earth. As an example, some 4000 NAV images were acquired on flight 4, but only 62 were retained.[7] The NAV images were later made into movies showing the path across the Martian landscape.

The NAV image acquisition rate was almost perfectly synchronized with the rotor spin rate, such that the orientation of the rotor blade shadows appeared roughly fixed from one frame to the next. Small differences in the rotor speeds, however, meant some rotation could be seen over the course of a flight.

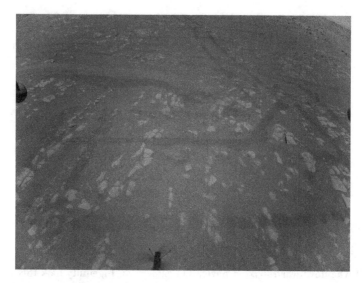

Fig. 6.11 This RTE camera image is the first color image taken by Ingenuity while aloft, during its second successful flight test, on 22 April 2021. At the time of this image, Ingenuity was 5.2 m above the surface and pitching (moving the camera's field of view upward) so the helicopter could begin its 7-ft (2-m) translation to the west, away from the rover. Rover tracks are apparent on the ground, and one of the landing feet intrudes into the field of view at left. The shadow of Ingenuity is seen at the bottom, with the shadows of the rotor blades rather blurred, because the exposure time of this camera is longer than the NAV camera. Source: NASA/JPL-Caltech.

Fig. 6.12 The Perseverance Mars rover is just visible in the upper left corner of this Ingenuity RTE image taken during its third flight, on 25 April 2021. The helicopter was 85 m from the rover at the time and at an altitude of 5 m. Sand ripples are more abundant in this scene than in the second flight. Notice the brightening of the image towards the center. Source: NASA/JPL-Caltech.

SOUND OF ROTORS ON MARS

Among the Perseverance rover's firsts was its recordings of sound on Mars (see Fig. 6.13). Microphones were installed on the cameras set to document the parachute inflation and landing, and on the SuperCam instrument, which used laser pulses to analyze the composition of rocks. The SuperCam microphone was used to listen to flights 4, 5, and 6; it was not used on the earlier flights because the simultaneous operation of the microphone with the Mastcam-Z camera and the helicopter base station had to be tested first. In later flights, the helicopter was too distant to be audible; beyond about 140 m distance, the attenuation of sound in the Mars atmosphere is too severe.[8] Specifically, carbon dioxide absorbs higher frequency sound more strongly: even though laboratory tests (e.g., Fig. 6.14) show that many harmonics of the blade crossing frequency (2 blades × 2537 rpm × 2 rotors in opposite directions = 84 Hz) are generated, only the 84 Hz and sometimes 168 Hz tones could be heard 50 m or more away on Mars.

In principle, in addition to the reception of radio signals from Ingenuity by the helicopter base station on the rover (see Fig. 6.15), a radar instrument (RIMFAX) on the rover may have the ability to detect Ingenuity, either with

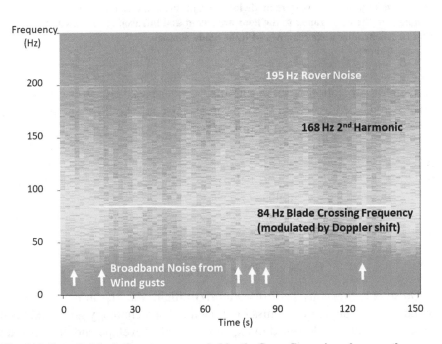

Fig. 6.13 Sound of the helicopter as recorded by the SuperCam microphone on the rover on Flight 4, shown as a spectrogram. The regular throbbing of the rotor blades at a near-constant frequency appears as a near-horizontal line, made slightly wavy due to the Doppler shift caused by the helicopter's motion away from and back toward the rover. Source: Author.

Fig. 6.14 The author at AeroVironment in September 2021, testing the acoustic emission from Terry, a terrestrial demonstrator unit, to help understand the sounds recorded on Mars. Although some tests were made in hover, in this instance the vehicle is secured by sandbags to the floor. Taped to the floor are sound and infrasound sensors and photodiodes to synchronize their readings with the blade passage.
Source: Author.

active radar operation or just using the receiver passively. However, testing would be needed to ensure such observations could be done safely, and no results had been obtained at the time of writing.

With the technology demonstration flights completed, the rover team was relieved of its responsibilities of photographing flights (see Figs 6.16 and 6.17). The geometries of future helicopter flights were guided in part by features of interest to the geologists on the rover team, to exploit Ingenuity's aerial view and ability to scout ahead of where the rover might drive. The helicopter's position would be determined by the data from its NAV camera.

Flight 6 Anomaly

On sol 91, Ingenuity performed its sixth flight, the first in the operations demonstration phase of the mission, now that the technology goals had been met. The flight was designed to expand the flight envelope and demonstrate aerial-imaging capabilities by taking stereo images of a region of interest to the west. Ingenuity was commanded to climb to 10-m (33 ft) altitude before translating 150 m to the southwest at a ground speed of 4 m/s (9 mph). Then, it was to translate 15 m to the south while taking images looking westward (see Fig 6.18), and then fly 50 m northeast and land.

Telemetry shows that the first 150-m leg of the flight was executed nominally, but then the helicopter started bucking violently, with roll and pitch excursions of more than 20 deg, large control inputs, and spikes in power consumption. Fortunately, it appeared to land safely, albeit several meters from the intended location.

What happened is that about 54 s into the flight, a glitch occurred in the pipeline of images being delivered by the navigation camera. This glitch caused a single frame to be dropped, but more importantly, it resulted in all later navigation images being delivered with inaccurate timestamps. Then, when the navigation algorithm derived a control correction based on what was seen in the image compared with what it predicted, the predictions were being made for the wrong time. The system's efforts to correct these phantom errors

Fig. 6.15 The monopole antenna of the Rover Base station, used to send commands to and receive data from Ingenuity up to 1 km away, is seen at the left of this image, casting a shadow on the rover deck. In the foreground is a camera target, with color patches for calibration. (This image, acquired on sol 2 by the Mastcam-Z camera, was intended to image the target before significant dust had accumulated on it.) The black-and-white circles help calibrate for the slow accumulation of Mars dust expected on the target, and the stalk is a sundial gnomon to cast a shadow to record the illumination geometry and isolate the direct solar illumination from that scattered by dust in the atmosphere. The square plate with polka-dot pattern is a set of targets for the SuperCam laser, and the multimission radioisotope thermoelectric generator (MMRTG) and its fins are visible at upper right.
Source: Cornell/NASA/JPL-Caltech.

resulted in the jerky control inputs. Fortunately, Ingenuity was programmed to ignore the navigation images during the last moments of descent (eg, in case the rotor downwash caused dust to be kicked up); working off the clean IMU data alone,‡‡ it stopped bucking around and landed with a level attitude.

The reason for the glitch was suspected to be the heavy central processing unit (CPU) load associated with acquiring and storing RTE color images. Thus, during flights 7 and 8 it was decided not to acquire any RTE images until a software patch for the Snapdragon was developed to avoid loss of synchronization of images with their timestamps. The update was implemented in time for the ninth flight. Meanwhile, in time for Flight 8, a software update was developed for the flight controllers, in order to prevent recurrence of the watchdog timer anomaly that had been an occasional nuisance earlier.

The helicopter team continued to push the envelope, demonstrating execution of flight characteristics beyond those in the original specification. For

Fig. 6.16 Two frames from Mastcam-Z, taken about 20 s apart during the fifth flight of Ingenuity, which flies in from the right (upper frame) and then climbs vertically (lower frame) to an altitude of 10 m, becoming clear of the horizon. It then descended vertically to a landing.
Source : Cornell/NASA/JPL-Caltech.

---

‡‡As Star Wars fans might put it, Ingenuity closed its eyes and used the force....

Fig. 6.17 Screen-grab of the author's computer running a 3-D visualization tool that lays orbital images over a digital elevation model to project a synthetic scene of Jezero from any chosen viewpoint. This shows a representation of the rover at its viewing position on sol 76; the white line shows the 129-m-long flight path of flight 5 of the helicopter (as imaged from the rover in Fig. 6.16). Note the vertical leg at the left end of the flight. A sampling system cover (bellypan) and the helicopter cover are just visible toward the bottom center of the image at a dead end of the rover tracks. This particular visualization is available on the web at https://www.perseverancerover.spatialstudieslab.org.
Source: Author.

example, flight 9 hopped across a sandy region Séítah,[§§] requiring a traverse of some 625 m (0.4 miles) (see Fig. 6.19).

Flight 10 targeted an area called Raised Ridges (RR), which had been identified by the Perseverance rover team as intriguing. This flight had a nominal (record) altitude of 12 m and required 10 distinct waypoints. On its way, the plan was to take several pairs of RTE images, each pair separated by a sideways flight segment to give a horizontal separation of the viewpoints. This would allow a stereoscopic analysis of the overlapping areas of the images to make a high-resolution topographic map.

Although the helicopter had outpaced the rover, as might be expected in the aerial scouting paradigm, in fact, the slower cadence of helicopter flights, coupled with the growing speed of the rover, led to the rover starting to catch up. (The rover didn't actually drive at a faster speed, but rather the distance it could drive in a day was limited by its autonomous odometry and hazard

---

[§§]In prelanding mapping of Jezero crater, the region that Perseverance turned out to land in had been named after features in the Canyon de Chelly region of Northern Arizona. Accordingly, many of the features identified on the ground after landing were given names from the Navajo language. *Séítah* means sandy.

Fig. 6.18 An RTE image from Ingenuity at an altitude of 10 m during its sixth flight, on 2 May 2021. Note that the color balance of the RTE images varies widely—the balance of visible and near-infrared light on Mars is a little different than on Earth, and the camera processing sometimes results in odd sky shading. It is in part to avoid such issues that the scientific cameras on Perseverance have calibration targets. This northwest-looking image shows the shelf of delta deposits that will be the target of Perseverance's sampling operations.
Source: NASA/JPL-Caltech.

Note that because of its mounting, the RTE field of view does not reach the horizontal. It was only because the vehicle was pitched upwards here that this compelling view of the horizon was captured.

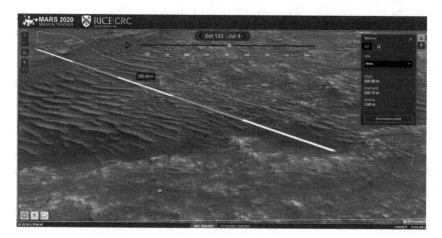

Fig. 6.19 Visualization of the record-breaking traverse of flight 9 from right to left (southward) across the tongue of sand ripples Séítah. The rover's position at the time is seen at lower left.
Source: Author.

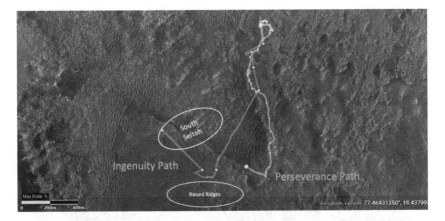

Fig. 6.20 This annotated HiRISE (High resolution Imaging science Experiment, a camera on Mars Reconnaissance Orbiter) image depicts the ground tracks of NASA's Perseverance rover and Ingenuity Mars helicopter since arriving on Mars. The upper yellow ellipse depicts the south Séítah region, which Ingenuity flew over during its 12th sortie.
Source: NASA/JPL-Caltech.

avoidance AutoNav software capabilities, which were progressively improved so that it could drive more minutes per day.)

Flight 12 was to acquire stereo pairs of images separated by 5-m side-step—10 RTE images in all to map the outcrops and ripples to help plan Perseverance's driving and scientific observations in south Séítah. This was ambitious because Ingenuity's navigation system relies on the assumption of flat terrain in correlating image features to derive an attitude and position solution. The more rugged terrain in south Séítah (see Figs. 6.20 and 6.21) leads to a stronger deviation from this assumption and hence noisier navigation data.

As the mission went on, the seasons on Mars changed. Although the amount of sunshine and the temperatures at Jezero changed only modestly, the changing illumination on the polar regions led to the accumulation of carbon dioxide frost (as happens every Mars year). This, in turn, depletes the atmosphere of this gas, and the atmospheric density declines.

Ingenuity had been designed to accomplish its five-flight mission within a few months of landing when prepared for flights at atmospheric densities between 0.0145 and 0.0185 kg/m^3, which is equivalent to 1.2%–1.5% of Earth's atmospheric density at sea level. However, six months after landing, densities were expected to drop to 0.012 kg/m^3 (1.0% of Earth's atmospheric density) during the afternoon hours preferred for flight.

Ingenuity was designed with a thrust margin of 30%, such that the rotors could provide 1.3 times the vehicle weight, allowing some extra to initiate climb and perform maneuvers. But if the atmospheric density were to drop to 0.012 kg/m^3 in the coming months, the thrust margin could drop to as low as

Fig. 6.21 RTE camera view from Flight 13. This image of an area the Mars Perseverance rover team calls Faillefeu was captured by NASA's Ingenuity Mars Helicopter on 4 September 2021. At the time the image was taken, Ingenuity was at an altitude of 26 ft (8 m). Images of this geologic feature were taken at the request of the Mars Perseverance rover science team, which was considering visiting the geologic feature during the first science campaign.
Source: NASA/JPL/Caltech.

8%, making the vehicle vulnerable to losing control or encountering blade stall in gusts.

Fortunately, a means to recover performance margin existed: spinning the rotors faster—to 2800 rpm instead of the nominal 2400 rpm (or maximum 2537 rpm operated on Mars to date). However, higher speed operation means higher blade frequencies with the possibility of exciting some undesirable resonance (which need not be structurally damaging to be a problem—if navigation sensors were just subject to excessive vibration, their performance could degrade). And higher rotor speed brings the tip Mach number higher, to about 0.8. This is probably still low enough to avoid shock wave formation, but again is closer to the edge of feasible performance. More electrical power is needed to achieve the higher speed, stressing the battery and electronics, and the mechanical loads on the rotor and swashplate components are larger. Thus, careful assessment of all the systems was needed to understand whether the flight could be performed safely in Mars conditions.

A high-speed rotor spin test on the ground was completed successfully on sol 204 of the Perseverance mission (September 15 or 16 on Earth, depending on the time zone). The motors spun the rotors up to 2800 rpm (with the collective pitch set to a low angle to avoid generating much lift), briefly held that speed, and then spun the rotors back down to a stop. Accelerometer data from the IMU showed that the higher spin rate did not unduly excite any resonant vibration modes in Ingenuity's structure. This ground test cleared the way for a planned test flight at 2700 rpm, executing a brief hover at 5-m altitude on sol 26. However, Ingenuity aborted the takeoff when it detected an anomaly in two of the small swashplate servo motors during its automatic preflight checkout.

During the automatic check (nicknamed a "servo wiggle"), Ingenuity commands each of the six servos (three for each rotor) through a series of steps over their range of motion and confirms that they reach the commanded position. The data indicated that two of the upper rotor swashplate servos—servos 1 and 2—began to oscillate through an angle of about 1 deg about their demanded positions just after the second step of the sequence. Ingenuity's software detected this oscillation and promptly aborted the self-test and flight.

Further wiggle tests to diagnose the problem were executed on sols 209 and 211, without showing the oscillation. One theory is that the high-speed spin test left the upper rotor in some unusual configuration that loaded servos 1 and 2 in a unique manner that drove the oscillation. Another theory is that mechanical wear had started to cause looseness in the servo gearboxes and swashplate linkages—after all, Ingenuity had made more flights and notched up more than double the design specification of flight-minutes at this point. Ground tests and analyses were needed to understand the problem, but time ran out, so Flight 14 could not be executed before Conjunction. Ingenuity's flights, summarized in Table 6.2, are also documented in the chief pilot's log book (Fig. 6.22).

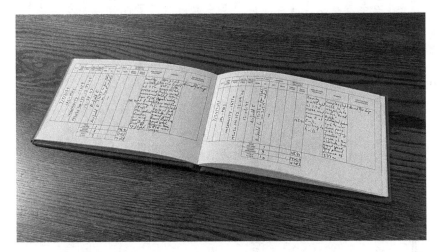

Fig. 6.22 Ingenuity logbook entries: Håvard Grip, chief pilot of NASA's Ingenuity Mars helicopter, documents the details of each flight in the mission's logbook, *The Nominal Pilot's Logbook for Planets and Moons*, after each flight. Entries for Flights 9 and 10 are seen here.
Source: NASA/JPL-Caltech.

TABLE 6.2 SUMMARY OF INGENUITY FLIGHTS THROUGH CONJUNCTION

Flight No.	Sol	Route	Summary
1	58	Hover at Wright Field.	The first powered flight by any aircraft on another planet. While hovering, it rotated in place 96 deg in a planned maneuver.
2	61	Hover, shift westward 2 m, hover, return, hover, land.	From its initial hover, it tilted 5 deg, allowing the rotors to fly it 2 m sideways. It stopped, hovered in place, and rotated counterclockwise, yawing from +90 deg to 0 deg to −90 deg to −180 deg in three steps, to point its color camera in various directions to take photos. It then returned to its takeoff point.
3	64	Hover, shift northward 50 m, return, hover, land.	This was the first flight to venture some distance from the helicopter's deployment spot. It flew downrange 50 m at a speed of 2 m/s. After a short hovering above the turnback point, it returned to land at the departure spot.
4	69 (attempt on sol 68 failed)	Hover, shift southward 133 m, hover, return, hover, land.	On this flight the helicopter took color images while hovering at its farthest point from takeoff. The Perseverance rover recorded the sound of the flight. In this flight, Ingenuity overtook Perseverance in the distance they traveled during the mission.
5	76	Hover, shift southward 129 m, climb to 10 m, hover, land at Airfield B.	This was the first flight to land at a new location 129 m south. On arrival, it gained altitude, hovered, captured a few color terrain images, and then landed at that new site, Airfield B. This flight was the last in the technology demo phase. The sound of the flight was recorded.
6	91	Translate southwest about 150 m, southward about 15 m, northeast about 50 m (160 ft), land near Airfield C.	This flight was the first in the operation demonstration phase. Toward the end of the first leg of the route, a glitch in the navigation image processing system led to jerking control. Ingenuity landed about 5 m (16 ft) away from the planned site, assumed as its Airfield C, at the time only previously observed in orbital imagery. The sound of the flight was recorded.
	107 (attempt on sol 105 failed)	Shift southward 106 m to land at Airfield D.	Ingenuity flew 106 m south to a new landing spot and landed at Airfield D. Use of the color camera was avoided to prevent recurrence of the flight 6 glitches.
8	121	Shift south southeast 160 m to land at Airfield E.	Ingenuity flew 160 m south to land at Airfield E, about 133.5 m away from Perseverance. As on the previous flight, the color camera was not used.

9	Shift southwest 625 m to Airfield F.	Ingenuity flew a record length of 625 m southwest, over Séítah, a prospective research location in Jezero crater, at a record speed of 5 m/s. This was a risky flight, straining the navigation system, which assumed flat ground whereas Séítah had uneven sand dunes. This was partly mitigated with the helicopter flying slower over the more challenging regions of the flight. Due to these errors, Ingenuity landed 47 m from the center of the 50-m radius airfield.
10	Loop south and west over Raised Ridges to Airfield G.	Ingenuity looped south and west over Raised Ridges, another prospective research location on Mars. Unlike the previous one, Perseverance planned to visit here. Ingenuity flew a total distance of 233 m (764 ft) past 10 waypoints, including takeoff and landing, at a record height of 12 m (39 ft).
11	Shift northwest 383 m to land at Airfield H.	This flight was primarily intended as a transition to a new takeoff point from where the next flight for the photographs of south Séítah region was planned.
12	Roundtrip northeast for 235 m, landed again at Airfield H.	The roundtrip was about 235 m northeast and back. The return path was laid about 5 m aside to allow another attempt of paired images collection for stereo imagery. As a result, the helicopter landed about 25 m east from the takeoff point.
13	Roundtrip northeast for about 105 m, landed again near Airfield H.	Explored ridge in Seitah

CONJUNCTION

When the sun is within a couple of degrees of Mars as seen from Earth, or vice versa, radio emission from the solar corona introduces a strong background noise into spacecraft communications links. Thus, routine operations must be suspended and the spacecraft must largely fend for themselves, operating on a prestored sequence. Operations that might be hazardous, such as driving or flying, are not attempted. Relatively simple passive data acquisition, such as imaging or meteorological monitoring, can continue if the memory storage on the lander or rover allows (or if those data can be offloaded to one of the spacecraft orbiting Mars—the solar conjunction does not affect those links).

The solar conjunction of September/October 2021 marked a suspension in Ingenuity operations and an appropriate milestone at which to close this story. (Perseverance halted driving operations, with a command moratorium beginning on sol 217, although some preprogrammed meteorological observations were expected to continue in the blind.) Ingenuity flights did resume after conjunction and were ongoing at the time of writing.

REFERENCES

1   Much of the material in this chapter is derived from the blog and status entries at the Mars Helicopter Technology demonstration webpages at https://mars.nasa.gov/technology/helicopter/. The Wikipedia compilation is also an excellent resource: https://en.wikipedia.org/wiki/Ingenuity_(helicopter). Another useful web location is the unmannedspaceflight.com forum, where much excellent amateur image processing work is reported.

2   Balaram, B., and Golombek, M., "Mars Helicopter Proposal Information Package," *Mars 2020 Participating Scientist program NNH19ZDA001N-M2020PSP*, https://nspires.nasaprs.com

3   Moorman, R. W., "Ingenuity Takes Off on Mars," *Vertiflite*, May/June 2021, pp. 16–21.

4   Grip, H. F., Lam, J., Bayard, D. S., Conway, D. T., Singh, G., Brockers, R., Delaune, J. H., Matthies, L. H., Malpica, C., Brown, T. L., and Jain, A., "Flight Control System for NASA's Mars Helicopter," *AIAA Scitech 2019 Forum*, AIAA-2019-1289.

5   The modes listed in the previous reference are slightly updated from those in an earlier development report: Grip, H. F., Scharf, D. P., Malpica, C., Johnson, W., Mandic, M., Singh, G., and Young, L. A., "Guidance and Control for a Mars Helicopter," *2018 AIAA Guidance, Navigation, and Control Conference*, AIAA-2018-1849. I have used information from both.

6   The Ingenuity project's blog on the various flights, and the resolution of problems, is at https://mars.nasa.gov/technology/helicopter/status/

7   Wikipedia, *Ingenuity (Helicopter)*, https://en.wikipedia.org/wiki/Ingenuity_(helicopter)

8   Maurice, S., Chide, B., Murdoch, N., Lorenz, R. D., Mimoun, D., Wiens, R. C., Stott, A., Jacob, X., Bertrand, T., Montmessin, F., Lanza, N. L., Alvarez-Llamas, C., Angel, S. M., Aung, M., Balaram, J., Beyssac, O., Cousin, A., Delory, G., Forni, O., Fouchet, T., Gasnault, O., Grip, H., Hecht, M., Hoffman, J., Laserna, J., Lasue, J., Maki, J., McClean, J., Meslin, P.-Y., Le Mouélic, S., Munguira, A., Newman, C. E., Rodríguez Manfredi, J. A., Moros, J., Ollila, A., Pilleri, P., Schröder, S., de la Torre Juárez, M., Tzanetos, T., Stack, K. M., Farley, K., Williford, K., and the SuperCam team. In Situ Recordings of Mars Soundscape, *Nature*, March 2022.

Chapter 7

DRAGONFLY: CONCEPTION AND DESIGN

"One of the first steps in any new endeavor is overcoming the discouraging advice of those who say it can't be done."

—Charles Kaman (1919–2011), helicopter pioneer

OVERALL CONCEPT

The concept of a rotorcraft lander trickle-charging a battery for brief atmospheric flights on Titan using the power from a radioisotope power source was outlined, academically, in 2000. At that time, the vehicle was imagined to be a helicopter. The concept was reimagined in early 2016 in response to the NASA New Frontiers community announcement.[1] Missions to Titan were solicited in the frame of "Ocean Worlds": "The Ocean Worlds theme for this announcement is tentatively focused on the search for signs of extant life and/ or characterizing the potential habitability of Titan or Enceladus"; thus, astrobiology would be a key focus of the mission.[2]

The key architectural considerations[3] that drove the mission design were that the vehicle would have to perform direct-to-Earth (DTE) communication, necessitating a high-gain antenna; it would have to access surface material for scientific analysis; it would have to be packaged inside an aeroshell for delivery to Titan; and it should operate using a single multimission radioisotope thermoelectric generator (MMRTG). These factors were never actually written down; the concept development at this stage was holistic. Indeed, they were never actually strict requirements—a vehicle with two MMRTGs is possible and would be more capable, but the MMRTG cost[4] was a significant (and initially uncertain) element of the overall, and very strict, Principal Investigator-managed cost cap of $850 million. This cost limit was a requirement that *was* written

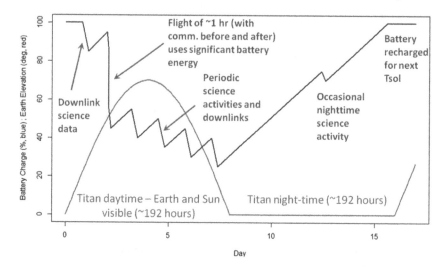

Fig. 7.1 **Energy management and communication concept of operations. The MMRTG continuously recharges the battery, but downlink and especially flight demand significant energy. Activities can be paced to match MMRTG in-situ capability while maintaining healthy margins on the battery state of charge.**
Source: Author.

down in the announcement of opportunity to which mission proposals would be submitted. Thus, although technically possible, it seemed that a two-MMRTG design would be unaffordable, and it was never seriously considered.

A brief evaluation by the author using a parametric rotorcraft flight power model indicated that a vehicle of representative size, powered by a single MMRTG (Fig. 7.1), could achieve unparalleled regional mobility on Titan, and the Dragonfly concept was born. Initially, it was imagined that the vehicle might have a flotation ring, to permit landing on one of Titan's lakes, but a more conventional box-with-skids layout soon emerged once it was decided that operations on dry land would be the focus of the mission.

A constraint in this application that is somewhat unusual for rotorcraft is the necessity to be packaged in a hypersonic aeroshell (see Fig. 7.2). The geometric tradeoff of unblocked rotor disk area vs number of rotors[5] with such a constraint suggests that, in fact, four rotors is the optimal number.

Although there is a small aerodynamic penalty in the "over-under quad" octocopter layout (with a top–bottom pair of motors/rotors at each corner of the vehicle; see Fig. 7.3) compared with a pure quad, the octocopter configuration is more failure-tolerant, being able to fly despite the loss of at least one rotor or motor. Because it was recognized that perceived risk would be high for a rotorcraft lander on Titan, the promise of resilience to failure was considered important enough to justify the additional hardware.

The geometric constraints on the vehicle posed by the aeroshell were significant, and the 3.7-m aeroshell itself had to be sized to fit inside a 4-m fairing

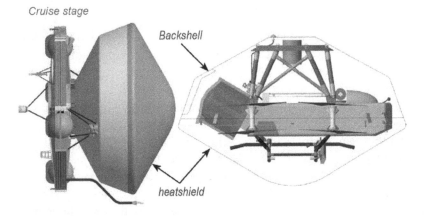

Fig. 7.2   Delivery in a Curiosity-like aeroshell defined geometric constraints that strongly drive the design of the Dragonfly vehicle. Center of mass considerations were important as well as the envelope.
Source: Applied Physics Laboratory (APL).

on the launch vehicle.* As well as defining the overall envelope of the lander, the center of mass of the aeroshell/lander assembly had to be sufficiently far forward to assure aerodynamic stability in the hypersonic entry phase of the mission. These factors drove the lander to be mounted on a strut assembly to meet the mass properties constraint and to exploit the maximum diameter of the aeroshell. In turn, the limited volume available forward of the lander meant the landing legs would have to be hinged and swing into place after the heat shield was released.

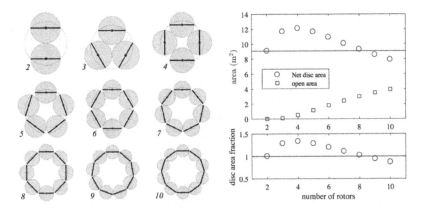

Fig. 7.3   Subject to the constraint of fitting in a circular perimeter, a quad-rotor is most efficient in terms of the unblocked disc area.
Source: J. Langelaan, by permission.

---

*Proposers to the New Frontiers program could elect to use a 5-m fairing if their mission demanded it, but doing so brought a cost penalty. Thus Dragonfly as proposed squeezed into a 4m fairing.

SCIENTIFIC PAYLOAD

The raison d'être of a lander vehicle in a mission of scientific exploration is to deliver and support a science payload. The centerpiece of the scientific payload of Dragonfly would be a mass spectrometer,[6] an instrument that can tease apart and identify the complex organic chemicals known to exist on Titan. The architecture of the sample acquisition system to bring solid surface materials into the mass spectrometer was another major early design choice. A sampling arm like those used on Viking or Phoenix, or indeed Perseverance, was briefly considered, but having many degrees of freedom, each with an actuator that would need to operate in cryogenic conditions, would be expensive and presented a number of single-point failures. Another key point is that reliable drill operation requires a force to be applied, the "weight on bit," and so the arm would have to be strong and thus heavy. Among the many aspects of Dragonfly, the natural design tension between a rotorcraft that seeks light weight and a drilling platform that seeks the opposite is a particularly novel and interesting one! The author has not been able to locate a previous instance of a helicopter with a rock drill.

Instead of an arm, two sample acquisition drills, one on each landing skid (Fig. 7.4), with simple 1-degree-of-freedom stages were chosen.[†] These provide the possibility of sample diversity at a given site and redundancy against failure, while using relatively simple, reliable mechanisms.

Titan's dense atmosphere permits the sample (whether sand, icy drill cuttings, or another material) to be conveyed pneumatically[7] through a tube by a blower. Pneumatic conveying is widely used in industries where powders or particulate materials are handled, such as the food industry, mining, or pharmaceuticals. A relatively simple valve arrangement would allow the selection of samples from the port or starboard drill, and the tubes would have a flexible section or a sliding joint to accommodate the landing leg deployment.

It was initially imagined that the sample would be introduced into the mass spectrometer using a cyclone separator, a common device in industrial settings (also familiar in Dyson vacuum cleaners). However, concerns about the possibility of sticky materials on Titan, and the weaker gravity on Titan, led to an innovative design where "tea strainer" sample cups would intercept part of the sample-laden airstream. These notionally single-use sample cups, which use a fine mesh to capture the sample but allow some air flow through, ensured that only the needed small amount of sample would be brought into the instrument, and it minimized any cross-contamination between successive samples.

[†]Whereas now, with the billion-dollar project in implementation and hundreds of people involved, major design changes require formal review and approval, in the fast-and-loose early days of the mission formulation by just a handful of zealots, this decision was basically made during an early 2016 phone call between myself and Kris Zacny at Honeybee Robotics.

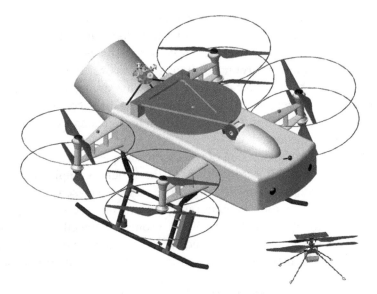

Fig. 7.4 The Dragonfly configuration as designed in the Step 1 proposal in April 2017. The blunt nose is shaped to fit in the aeroshell: no effort at streamlining had been made at this stage for atmospheric flight. Note the aerodynamic fairing in front of the gimbal for the grey circular HGA stowed flat. The cylinder at rear is the MMRTG. A sampling drill mechanism is visible in the nearside skid leg, and forward-looking cameras are recessed into the yellow insulating foam forming the rounded nose of the vehicle. Ingenuity is shown for scale.
Source: Adapted from APL graphic.

Like virtually every planetary mission, Dragonfly would have a suite of cameras. These would be needed to determine where to find the samples of most scientific interest and to document their geological context (eg, whether the material is dune sand, impact melt sheet, cobbles in a riverbed, etc.). Forward cameras and downward-looking cameras would be used for reconnaissance in flight, and on the ground these cameras would give an initial impression of each new landing site. A more detailed and comprehensive view of the landing site would be built up by a pair of pointable panorama[‡] cameras. These would be pointed by the two-axis gimbal mechanism already needed to point the high gain antenna (HGA), and so are mounted at its top to get the best view (Fig. 7.5). The use of two cameras of each type would give some redundancy against failure and also provided the opportunity to perform stereo imaging to measure topography at a range of scales. Although the other cameras are embedded in the warm body of the lander, these panorama imagers would need heating when they were operated, and so they have insulating housings.

[‡]The cameras are used to make panoramas by creating a mosaic of a set of images from different pointings. The cameras are not themselves wide-field (panoramic) imagers.

An additional pair of cameras is mounted in the lander belly, with a lens system to provide closeup views (resolving individual sand grains) of the surface material. In addition to providing geological context information, such as whether the grains are rounded indicating erosive transport from their source area, these microscopic imagers—somewhat similar to the "hand lens" imager on Mars rovers but with a longer stand-off distance—would indicate the texture of the surface material and inform the decision as to whether the material should be sampled.

Another factor in the decision of whether to sample at a given landing site is the overall elemental composition: Is the ground completely made of organic hydrocarbons and nitriles, or is there water ice present? This classification can be done with a gamma ray spectrometer (see Fig. 7.6), which reports the elemental abundances of the ground averaged over a volume roughly in a hemisphere under the lander. One nice feature of the Titan environment is that the high-resolution germanium detectors preferred for such

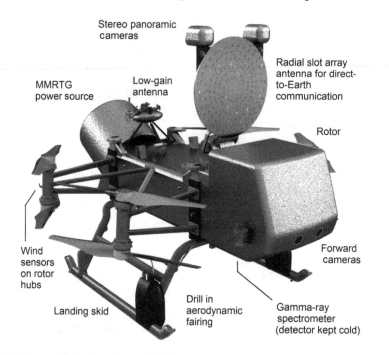

Fig. 7.5 **The configuration in April 2019, at the end of the Phase A study. Although the overall configuration is similar to that in Step 1, a few more details (and better graphics) are evident. One notable evolution is the "attic"—the nose section grew upwards to accommodate the drill for the acquisition of complex organics (DrACO) sample carousel and mass spectrometer. A wind sensor is mounted on each rotor hub, so that one is always on the upwind side of the lander. The panorama cameras on the HGA are now equipped with thick insulated housings to minimize the energy needed to warm them to operating temperature.**
Source: APL graphic, annotations by author.

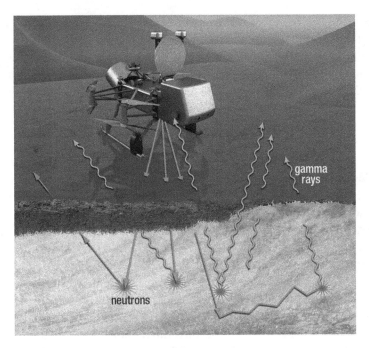

Fig. 7.6 Schematic of the operation of the gamma-ray and neutron spectrometer operation. A pulsed neutron generator bombards the surface with neutrons, some of which are scattered back to detectors, diagnosing the hydrogen content of the ground. Others stimulate atoms in the surface to emit gamma rays that have energies that fingerprint different elements in the ground.
Source: APL.

measurements need to be at temperatures below 100 K to operate, and in normal space missions typically must be equipped with cryocoolers (effectively, small Stirling-cycle refrigerators) to do so. However, on Titan such cooling is not needed; the detector can simply be sited outside the lander body, where it quickly attains Titan ambient temperature. Such exploitation of the Titan environment was a neat and novel soundbite to communicate early in the mission's development. On the other hand, the thick atmosphere on Titan (with its column mass equivalent to that of a 100-m layer of water on Earth) screens out almost all cosmic rays. It is these cosmic rays on Mars or airless worlds that excite atoms in the rocks to produce the diagnostic gamma rays. Thus, Dragonfly needed to carry a pulsed neutron generator (PNG), a device used in borehole instrumentation for oil exploration on Earth, to provide that excitation.[§]

[§]Although the MMRTG produces a flux of neutrons, which have to be taken into account in the design of the lander electronics, these neutrons are mostly at energies too low (<2 MeV) to excite diagnostic gamma rays. The PNG is a high-voltage tube, provided by the exploration services company Schlumberger, that accelerates tritium and deuterium atoms/ions to collide, yielding neutrons with an energy of some 13.6 MeV. It can thus be stated that Dragonfly is, in fact, equipped with a fusion reactor!

The fifth component of the payload is an environmental sensor package called DraGMet (Dragonfly geophysics and meteorology). This measures atmospheric conditions such as wind and methane humidity to inform about the aeolian and fluvial processes that may have redistributed Titan's surface materials. The package uses a single data processing unit to interrogate a range of sensors. Since the announcement of opportunity (AO) called for investigation of habitable environments on Titan, it was important to address the scientific question that bears on icy moons/ocean worlds generally, namely, how thick is the ice crust that overlies Titan's suspected internal water ocean? DraGMet answers this in a couple of ways. First, electric field measurements by the Huygens probe had suggested a possible Schumann resonance in the cavity formed by the ocean and the ionosphere[8]; thus, DraGMet has a set of electrodes to repeat these electric field measurements under a range of times of day and under much quieter conditions than the Huygens descent.

Second, and more importantly, if Titan should be seismically active, then seismic waves will reverberate in the ice crust (thought to be between about 50 km and 200 km thick), allowing its thickness to be measured by the propagation time of the waves through it. (It is likely that Titan is seismically active because it is in a slightly elliptical orbit around Saturn; thus, Saturn's tidal pull deforms the crust once per Titan day. The Apollo seismometers on our own moon detected a tidal variation in moonquake activity.) Hence, DraGMet incorporated a seismometer, lowered to the ground from the lander belly to minimize noise from lander equipment or shaking by the wind.

Thus, a payload that was responsive to the AO was conceived (see Table 7.1). Some further details on these instruments and how they are used can be found in the next chapter. One can imagine many other instruments that would be interesting to also include (a ground-penetrating radar is one that springs to mind), but the tight fiscal envelope of the New Frontiers AO, and the premium on the data volume that can be downlinked to Earth, precluded a more expansive payload. In fact, the instrument complement has not changed since its original formulation (largely by this author) in early 2016.

Dragonfly's drill and sample transfer system (Fig. 7.7) is being developed by Honeybee Robotics of Pasadena, California. The robust drill itself was evolved from that used by the Apollo astronauts on the moon, with a specially designed conical drill bit for sampling Titan's (softer) materials while preventing it from getting stuck.

VEHICLE DESIGN

In terms of functional requirements, the vehicle must convey the scientific payload to sites of interest and provide the payload with the services needed for its operation, namely electrical power, thermal management, and command and data handling.

TABLE 7.1   DRAGONFLY'S SCIENTIFIC PAYLOAD

Instrument	Lead	Technique
Drill for the acquisition of complex organics (DrACO)	Honeybee Robotics	Dual rotary-percussive drills with custom bit Pneumatic sample transfer system for rapid, cold transfer of drill cuttings to DraMS sample carousel
Dragonfly mass spectrometer (DraMS)	NASA Goddard Space Flight Center	Mass spectrometer with laser desorption and thermal/gas chromatography front ends to analyze molecular composition of ices and organic compounds in sampled material
Dragonfly gamma-ray and neutron spectrometer (DraGNS)	Johns Hopkins APL	Pulsed neutron generator with gamma-ray and neutron spectrometers to quickly measure elemental composition (carbon, nitrogen, hydrogen, oxygen, salts, etc.) of surface under lander.
Dragonfly camera suite (DragonCam)	Malin Space Science Systems	Forward and down-looking wide-angle cameras Pointable cameras for panoramas; close-up imagers to view drill sites and surface material LED illuminators for color information and organic fluorescence detection
Dragonfly geophysics and meteorology package (DraGMet)	Johns Hopkins APL	Wind, pressure, temperature measurement, and sensors for methane humidity and hydrogen gas Electric field and physical properties of surface Seismometer

As discussed in Chapter 1, the power source (assumed to be one MMRTG) is central to the mission concept, because not only is the electrical power required for operation on Titan's surface, but the "waste" heat also is essential in the Titan environment. A radioisotope power system (RPS) is therefore an enabling element for sustained Titan surface operations.

Like the Mars rovers Curiosity and Perseverance, heat is brought into the lander body using a pumped fluid loop, to sustain benign conditions for the battery and the lander electronics. Like the Huygens probe, Dragonfly features a thick foam insulation layer all around its body to limit the heat needed to maintain an internal temperature about 180°C (356°F) warmer than the outside. There are a few systems whose operation requires close coupling to the Titan environment, namely the drills, the rotor motors, and the cameras on the HGA gimbal. These are heated electrically when they are needed (which is only a small fraction of the time) and insulated as appropriate to minimize that heating energy.

One aspect of spacecraft design that is not widely appreciated is that landers actually have rather low electrical energy needs compared with deep-space vehicles or satellites. Indeed, the highly successful Viking landers (which, after all, had rather similar scientific goals to Dragonfly—astrobiologically

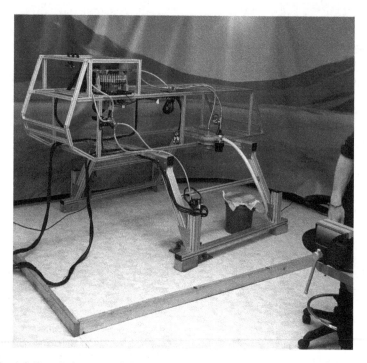

Fig. 7.7 A full-scale frame model of Dragonfly (the "Iron Bird") used in Phase A to test an all-up prototype of its sampling system, with drill, carousel, transfer hoses, diverter valves, and blowers.
Source: Author, at Honeybee Robotics.

motivated surface chemistry with imaging, meteorology, and seismic studies) operated—with 1970s electronics—on less than 70 W, supplied by two small RTGs. The key is that the lander can mostly be in a quiescent low-power state, where there are few direct loads and the RTG output can be accumulated in a battery. This is in contrast to a spacecraft in deep space, where it is usually obligatory to maintain continuous knowledge of its orientation [requiring inertial measurement units (IMUs) to be powered] and where hydrazine rocket propellant (which has a fairly high freezing point) requires that tanks, pipework, and valves be kept warm.

Although sample acquisition and chemical analysis are somewhat power-hungry activities, they require only a few hours of activity at each landing site. The seismic monitoring, however, has to be continuous, and so the DraGMet instrument is designed with particular attention to low-power operation—it runs in the background when the rest of the lander electronics are turned off to maximize the rate of recharge of the battery.

This hibernate/recharge concept of operations (CONOPS) was seen as an important factor in minimizing any perceived risk associated with degradation of the MMRTG output. Although when Dragonfly was conceived there were four years of operational history of the MMRTG on the Curiosity rover (where a power output loss of around 3%/yr has been seen[9]), and the decay of the ^{238}Pu output is deterministic, the thermoelectric converter degradation is temperature- and current-dependent and so may be slightly different for Dragonfly. However, a decline in electrical power does not fundamentally change Dragonfly's capabilities; it only requires a bit longer charging time between flights or downlinks. Thus, the mission can gracefully accommodate any degradation in power output. Ultimately, the heat output will become too low to maintain operating temperatures, but this is not expected for years after Dragonfly's nominal mission.

The architectural decision within the proposal framework to use a single MMRTG effectively set the overall energy budget for the mission, although the battery capacity is in principle a free design parameter. The energy budget, in turn, and the architectural requirement to perform DTE communication, basically determined the capability of the communication system, in that the antenna size in practical terms would be limited to the lander body width.

Although shorter wavelengths (Ka-band) are becoming preferred for interplanetary missions, in fact X-band (3 cm, 8 GHz) is preferred for Titan surface operation, because especially at low Earth elevations above the horizon, there is some appreciable attenuation of Ka-band signals in the dense Titan atmosphere.

Missions with HGAs empirically[10] require about 5 mJ per bit per astronomical unit (10 AU) to acquire and send science data to Earth. [The linear distance dependence is an interestingly emergent (allometric) correlation that results from engineering efforts to defeat the inverse square law—spacecraft at greater distances tend to have larger antennas, for example.] Thus, at Titan's typical distance of $\sim$10 AU, about 20 bits of science data can be sent to Earth for every Joule of energy (or equivalently, about 9 kilobytes/W· h).

A mission following on from Huygens should logically do better than it. The Huygens probe returned about 100 Mbit of data. To do, say, 100 times better (10 Gbit) would therefore require at 10 AU about 0.5 GJ of energy (140 kW · h, far beyond the capability of practical stored energy systems like primary batteries). The free parameter in the system design is the mission duration. For the steady output from a radioisotope power source, the mission energy, and thus data return, scales directly with duration. One year of (say) 100-W output corresponds to 3 GJ of energy.

It should be borne in mind that comparing raw data volumes with Huygens does not capture the science value of those bits. Huygens, being a one-shot probe, was obliged to hose out its scientific data essentially as fast as it acquired it, and the data compression technology of the time was modest in sophistication. With much more computing capability to perform compression and selection, and the ability to retain data on board, the science value per bit can be enhanced considerably. For example (as done on Mars rover missions), a large suite of highly compressed thumbnail images can be sent to the ground, and transmission of the full resolution data with minimal compression loss can be restricted only to the most interesting-looking of the images, as determined by the science team on the ground. Similarly a combination of on-board detection and ground analysis can focus on the most information-rich periods of meteorological or seismic data. In any case, it was obvious that with a years-long mission, the DTE data return from Titan would support ambitious scientific goals, despite the much lower data volumes than those that have become possible at Mars with its fleet of relay orbiters (100s of Gbit per day!).

MOBILITY

An informal guide to determining the vehicle capability[11] in early development was the mantra that it should be able to fly in one hop further than any Mars rover has driven in a decade (ie, about 40 km or 25 miles). In practice this was never an actual requirement, and indeed the final capability of Dragonfly will be somewhat less, but early in development it served to set expectations—that the mobility afforded by rotor flight would be transformative, unlike anything seen in planetary exploration to date.

At low latitude on Titan, the operations of a lander with DTE communication are paced by the diurnal cycle. A Titan solar day (Tsol) is 384 h long (16 Earth days). Seen from Titan, Earth in the sky is always within 6 deg of the sun. Interaction with Earth, and logically any operations requiring real-time observation (such as atmospheric flight), occur during the day; nighttime activities are generally limited, and power can be devoted to recharging the battery. Thus, a logical maximum size of the battery is that which completely captures MMRTG power (roughly 100 W, less later in the mission) during the Titan night, or about 200 h (Fig. 7.1). Thus, a representative battery size might be 14–20 kWh. Such a battery—coincidentally the capacity of the Apollo lunar rovers' batteries, and about a quarter of the size of the battery in a Tesla electric car—would be rather massive (140 kg or 310 lbs), assuming a representative specific energy metric for space-qualified batteries of 100 W · h/kg. In practice, a smaller battery could be chosen, sacrificing some energy-harvesting efficiency for lower mass and cost. At the very least, however, it would be essential to have enough energy to fly several kilometers in a single flight, to allow "hopping" from one interdune plain to another. The characteristic separation of Titan's sand dunes is about 4 km. Evaluation of the different power demands of flight as a function of flight speed suggested in early analysis (Fig. 7.8) an optimum (max range) speed of about 10 m/s (20 mph). Thus if the vehicle can fly for ~30 minutes, it can attain a distance of ~18 km (~11 miles).

As discussed in Chapter 2, Titan's atmosphere is both denser (4.4 times) and colder (94 K) than Earth's. The composition is predominantly (95%) nitrogen, and the low temperature means molecular viscosity is rather lower

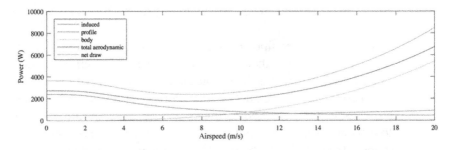

Fig. 7.8 **Rotorcraft power curve for a representative vehicle mass of 420 kg on Titan.** (This was a very early reference case using the mass of the Phoenix lander on Mars. The actual mass of Dragonfly may be around double this value.) The induced power required for rotor thrust falls toward higher speed, whereas the body drag increases quadratically and eventually dominates. These competing factors define the maximum endurance speed (the minimum in the curve~8 m/s) and the maximum range speed (where the tangent to the curve passes through the origin, corresponding to ~10 m/s). Titan's dense atmosphere and low gravity mean that the flight power for a given mass is a factor of more than 40 times lower than on Earth.

than for our air. The combination of higher density and lower viscosity means that an airfoil of given size and speed is operating at a Reynolds number that is several times higher than on Earth. To a first order, then, the ~1-m rotors of Dragonfly should resemble rotors of much larger scale systems on Earth—in fact, a blade section more typically used in terrestrial wind turbines has been adopted. (Rotor experts at Penn State University are a key part of the Dragonfly team.) This section is not only aerodynamically efficient, but also very tolerant of surface roughening (typically, in the case of wind turbines, due to insect impingement), making it a robust choice for Titan. Although the sound speed is lower in Titan's atmosphere than Earth's (~195 m/s vs ~340 m/s), the tip Mach number is still quite modest.

In addition to horizontal mobility, there is scientific value in achieving altitude. Of particular interest is the possibility of profiling the planetary boundary layer (PBL) via ascent to ~3.5-km altitude. The diurnal PBL thickness was measured during the Huygens descent to be ~300 m high, although a possible feature at ~3 km has been identified and attributed to a possible seasonal PBL; it is this quantity that apparently controls the spacing of dunes on Earth and Titan.

Although vertical ascent is possible, vertical descent is not (except at very low speeds, as for landing) because the vortex ring state, wherein the vehicle falls through its own downwash creating an unstable condition, must be avoided. Descending vertically at very low speeds would also be very energy-inefficient. Nominally, then, profiling flights (Fig. 7.9) would be performed with normal forward motion, ascending or descending at about 20 deg to the horizontal. These could be performed during traverses to new locations; or, if it was desired to make a local vertical profile with minimal horizontal displacement, a spiral ascent and descent could be executed with return to the original landing site.

A substantial amount of equipment is needed to support the mobility function. In addition to the rotors themselves (Fig. 7.10), the electrical motors driving them are not insubstantial. They must exclude dust, survive storage at cryogenic temperatures, and then deliver the appropriate torque and speed to permit the rotor speed to be varied for maneuvering. This, in turn, requires several kilowatts of power to be supplied to each motor, which is on the end of an arm spacing it from the lander body. Having the required thickness of copper wire becomes a significant mass consideration. This, in turn, motivated the use of an electrical bus voltage higher than the standard 28 V used on spacecraft[¶]; in fact, Dragonfly uses 100 V. The motor drive electronics are similarly formidable—pushing 10–20 kW of total power to eight different motors with possibly rapid changes in the power to each is not trivial.

But the most salient challenge, at least as far as superficial evaluation is concerned (and the early development of a competitive mission proposal is subject to evaluation by reviewers[12] whose expertise is unlikely to coincide directly with the topic at hand) is of autonomous flight, and in particular, landing.

In the Dragonfly development, extensive use was made of experience at APL in the development of Mighty Eagle, a prototype lunar lander.[13] Although

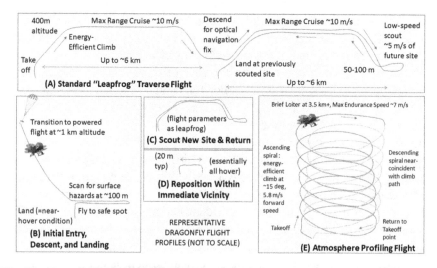

Fig. 7.9   Dragonfly flight profiles.

¶The 28 V is effectively a spacecraft inheritance from the aircraft industry, which had generally higher power demands than the automotive industry from which it drew hardware as equipment became standardized in the 1920s to 1940s. Auto electrics were set by the 12-V characteristic of six-cell lead-acid batteries, which in turn demanded generators that put out about 14 V to charge them. Because flights of many hours rely more on generator capacity than on the battery itself, the aircraft industry evolved to use 28 V as the nominal voltage. Even though battery chemistries and thus cell voltages are now typically quite different spacecraft tend to use 28 V, too.

Fig. 7.10   The Phase A design rotor, milled from aluminum alloy, held by the Dragonfly Principal Investigator, Dr. Elizabeth Turtle. Although other materials may be lighter weight (and may be pursued as the design develops), a one-piece alloy construction eliminated any concern in the proposal stage that the cryogenic conditions on Titan would cause material problems.
Source: Author.

this was obviously a rocket-powered vehicle, the guidance and navigation challenges are very similar: The vehicle must control its horizontal and vertical velocity to be near-zero at zero altitude.

A separate challenge is that of hazard detection. The vehicle should not attempt to touch down at a location where either the slope is excessive or small-scale terrain height variations due to rocks or gullies would result in adverse contact of vehicle components with the ground. Somewhat by tradition, allowable slopes (for both terrestrial rotorcraft and planetary landers) are assumed to be 10 deg. A similarly quasi-arbitrary specification of a 25-cm-high (50-cm-diameter) rock has been adopted.

Detection and avoidance of excessive slopes or rock hazards is implemented with a lidar instrument. Essentially, Dragonfly flies along a swath of terrain and scans it with lidar.[14] Areas of steep slopes and any obstacles (points with surface height >25 cm (0.8 ft) above their surrounds) are masked out, and safe landing areas are defined as the largest patches in this map that are free of these hazards (see Fig. 7.11).

This somewhat limited degree of autonomy (it cannot be described as artificial intelligence) assures the ability to find a safe landing spot should one exist, and the following chapter explains the basis for expecting that such spots should. In the Dragonfly mission as proposed, it was assumed that the lidar would be supplemented by inertial navigation, an optical navigation system (using dedicated down-looking navigation cameras, electronically and optically similar to the down-looking science cameras but with some implementation differences), and radar to measure altitude and velocity as on previous Mars landers (see Fig. 7.12).

It was recognized that the rotor downwash might kick up dust to cause brownout[15] in the last $\sim$10 m of descent, and thus lidar and camera information is discarded just prior to landing. It was assumed that an ultrasonic altimeter (used on terrestrial drones—the Huygens probe had an acoustic sounder that detected sound echoes from the Titan surface from an altitude of about 100 m) would be used to enhance the precision of altitude knowledge, and thus minimize the vertical velocity at ground contact, just prior to landing.

As on Earth, soft and hard terrain can be expected. The heaviest structural loads derive from hard terrain, which can be assumed to be completely unyielding, and thus all the kinetic energy of landing must be absorbed by the lander structure. This is a problem that confronts all planetary landers; the most mass-efficient solution is typically a component that absorbs energy by permanent deformation (eg, the Apollo lunar landers had crushable aluminum honeycomb inserts in their landing legs). This approach, however, is not applicable to a vehicle that makes multiple landings. Thus, pneumatic dampers were assumed for Dragonfly's landing legs. Under the compressive load of landing on a hard surface, these would dissipate energy by a piston forcing air

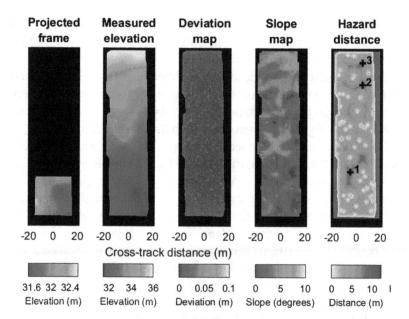

Fig. 7.11 Autonomous hazard avoidance uses lidar to measure surface heights in a square patch under the vehicle. As the vehicle flies, these frames are combined to make an elevation map from which local height deviations (eg, rocks or gullies) and slopes are computed, and safe landing sites are identified (right) that maximize the distance from these hazards. Source: APL.

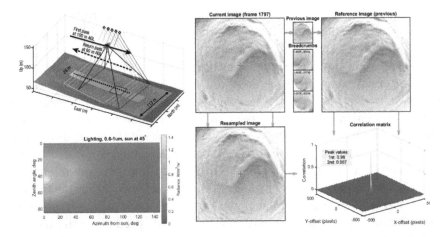

Fig. 7.12 The optical navigation system functions somewhat analogously to that on Ingenuity, correlating features in down-looking images. Rigorous simulations demonstrated that such an approach would work, despite the somewhat diffuse illumination on Titan. Source: APL.

through a small orifice. Upon a subsequent takeoff, a spring would re-extend the piston, refilling the damper cylinder with Titan air.

A separate challenge from the guidance and control aspects of mobility is the navigation function. The decision of where Dragonfly should go would be made (essentially specified as north/east coordinate offset from the current landing site) by the science team on the ground, based on the mission objectives and all the available terrain data (Cassini, and aerial imaging from Dragonfly). However, once flights of more than a few hundred meters are considered, the expected drift of the IMU would prevent the vehicle from reliably achieving the needed position accuracy. Thus, optical navigation would be employed.[16]

So-called "breadcrumb" images would be used as position references. The ability to correlate images of the terrain—so natural for human pilots—places some demands on the navigation system. The images must be similar enough to permit automatic registration of features within a fraction of a second. In practical terms, this means the images should be acquired at the same time of day, so that the illumination is similar. The altitude and orientation should be the same, or at least known so that the appropriate geometric transformations can be made. Dedicated processing hardware, in addition to the navigation cameras themselves, would be installed.

ENVIRONMENT SPECIFICATION

The deep interactions of a vehicle like Dragonfly with the environment[17] demand that a large number of parameters of the atmosphere and surface be specified for engineering design. In some cases, these are quite well-determined by measurements from the Huygens probe; in other cases, scientific judgement must be applied, with physical reasoning, or terrestrial analogs, applied to assert ranges that conservatively embrace the possibilities at Titan, desirably without being overconservative.

Among areas where the Huygens measurements give direct constraints on the Titan environment are the atmospheric properties (which were in any case rather well-estimated from Voyager data) such as pressure and temperature. Less well-constrained, due to their intrinsic variability, are winds. The Huygens descent yielded a snapshot profile of the wind speed and direction at about 0945 local time, at a site at 10 deg south, at a particular season. We must rely—as do Mars missions—on models to estimate how the winds might vary. A study[18] of the prevailing (global-scale) winds conducted during the Phase A study showed that three different global circulation models (GCMs) predicted the same direction (from the north-northeast) and speed (typically 0.4 m/s, and no more than 1.0 m/s or 2 mph), about the same as those measured by Doppler tracking of the Huygens probe. Thus, the 10-m/s flight transit speed means that wind effects on range are minor.

A more detailed evaluation of winds has been necessary to answer various design questions. For example, the wind chill that affects the thermal design (eg, how much heater power may be needed for the drill motors) depends on sustained winds. But questions of topple stability against wind loads (with the high-gain antenna deployed) demand consideration of the extreme gusts that might be encountered, even if such gusts last only a few seconds, driven perhaps by convective turbulence like dust devils. Thus, including a contribution for slope winds, and convection, led to a specification of 2 m/s for maximum sustained winds and 4.6 m/s for the maximum gust.[19] Specifications of turbulence are needed for the flight control system and for considerations such as wind noise on the seismometer.[20]

The properties of the surface are important. Among these, the mechanical properties are relevant for the design of DrACO, as well as for sizing the landing skids to avoid sinking too deep into soft terrain. An instrument on the Huygens probe, a penetrometer, measured the mechanical resistance of the Titan surface to be about 150 kPa (21 psi), but of course this was at only one spot in a landscape that we know is diverse in both geomorphology and composition.[21] As in the design of other planetary landers, rational specifications have to be adopted for allowable slopes and obstacles (rock sizes) and surface friction. The fact that Dragonfly will take off again means the unusual question arises[22] of how much adhesion might be developed by the ground on the skids.**

**A value of 3 kPa has been provisionally adopted. Because dealing with mud is such an ancient and empirical challenge, there is actually rather little quantitative treatment of this problem in aviation. The food industry frequently deals with the properties of organic slurries, and the 3-kPa value is equivalent to "sticky" wheat dough.

There are many environmental details that are not problematic[††] but need to be documented to show that they are not problematic. As an example, lidar systems tend to operate at only one or two specific wavelengths of light—those associated with the lasing medium that generates the light pulses. A popular one is 1.064 μm, associated with a neodymium-doped yttrium aluminum garnet (mercifully abbreviated to Nd-YAG). Another wavelength is 0.905 μm, generated by semiconductor laser diodes. Both wavelengths work well on Earth; however, the Titan atmosphere is 5% methane, and methane gas has a number of characteristic absorption bands—wavelengths at which the molecule absorbs energy. It was through these absorption bands in the red[‡‡] and near-infrared that Titan's atmosphere was discovered in 1944.

The 1.064 μm wavelength is nicely in between some of these methane bands, and so this lidar works well on Titan. However, 0.905 μm is rather close to the 0.889 μm methane absorption band,[23] and results in substantial attenuation in the Titan atmosphere, so this wavelength cannot be used.

Another set of risks, of concern in terrestrial aviation, had to be considered. These pertain to methane condensation. Could methane form ice on the rotor blades, as supercooled drops of water do in terrestrial storm clouds? The answer is no—the abundant nitrogen in Titan's atmosphere forms a solution with methane, depressing its freezing point well below the surface

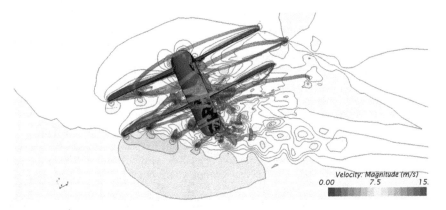

Velocity: Magnitude (m/s)
0.00   7.5   15.

Fig. 7.13 Computational fluid dynamics simulation of the flowfield around the Dragonfly rotors in forward flight. The shading denotes the flowspeed: The grey tubes are the tip vortices in which methane condensation could occur. The vortices dissipate quickly, so any condensation would not persist long enough to be advected under the lander where it could interfere with optical sensing.
Source: M. Kinzel.

[††] I have formulated a law on this topic: "It is always possible to contrive, with a probability that cannot be shown to be zero, a pathological planetary environmental scenario that will defeat any finite-cost space system." You can never say never.

[‡‡] It is largely due to the red absorption by methane that the planet Neptune looks blue. Water has similar absorptions, which is why illumination in the sea looks blue.

temperature.[24] And in any case, Dragonfly will be operating at a latitude and season where storms or even clouds were observed by Cassini not to occur, and GCMs confirm they should not form, so the vehicle should not encounter condensed methane. Could methane fog form in the tip vortices from the rotors (Fig. 7.13), as fog forms over the wings and in tip vortices of airliners on moist days on Earth? And might such fog give false triggers to the lidar? Again, the answer is no. Such condensation could occur[25] in stressing conditions, like the high-altitude portion of a profiling flight, but the low-pressure/temperature zone of the tip vortex is limited to within a meter or so of the rotor disk. Could camera windows get cold during profiling flights and cause methane condensation on their surface, like the fogging of glasses walking into the warm, moist air of a supermarket from the cold parking lot in winter? Once again, the answer is no. An evaluation of the heat leaking through the camera windows suggests they will always be warm enough to inhibit condensation. The list of such questions goes on and on (and so it should—better to spend time retiring a hundred phantom risks than to let one real risk go unnoticed).

References

1   Planetary News, *New Frontiers AO Community Announcement*, 2016, https://www.lpi.usra. edu/planetary_news/2016/01/08/new-frontiers-ao-community-announcement/. The surprise announcement on 6 Jan. 2016 indicated an intended release date for the draft announcement of opportunity in July 2016 and the final AO in Jan. 2017 with proposals due 90 days after that. The mission target list was: Comet Surface Sample Return, Lunar South Pole–Aitken Basin Sample Return, Ocean Worlds (Titan and Enceladus), Saturn Probe, Trojan Tour and Rendezvous, and Venus In Situ Explorer.
2   See, for example, Barnes, J. W., E. P. Turtle, M. G. Trainer, R. D. Lorenz, S. M. MacKenzie, W. B. Brinckerhoff, M. L. Cable, C. M. Ernst, C. Freissinet, K. P. Hand, A. G. Hayes, S. M. Hörst, J. R. Johnson, E. Karkoschka, D. J. Lawrence, A. Le Gall, J. M. Lora, C. P. McKay, R. S. Miller, S. Murchie, C. D. Neish, C. Newman, J. Núñez, M. Panning, A. Parsons, P. Peplowski, L. Quick, J. Radebaugh, S. Rafkin, H. Shiraishi, J. Soderblom, K. Sotzen, A. Stickle, E. R. Stofan, C. Szopa, T. Tokano, T. Wagner, C. Wilson, R. Yingst, K. Zacny and S. Stähler, "Science Goals and Objectives for the Dragonfly Titan Rotorcraft Relocatable Lander," *Planetary Science Journal*, Vol. 2, No. 4, p. 130, https://iopscience. iop.org/article/10.3847/PSJ/abfdcf/meta
3   Lorenz, R. D., E. P. Turtle, J. W. Barnes, M. G. Trainer, D. S. Adams, K. E. Hibbard, C. Z. Sheldon, K. Zacny, P. N. Peplowski, D. J. Lawrence, M. A. Ravine, T. G. McGee, K. S. Sotzen, S. M. MacKenzie, J. W. Langelaan, S. Schmitz, L. S. Wolfarth, and P. Bedini., "Dragonfly: A Rotorcraft Lander Concept for Scientific Exploration at Titan," *Johns Hopkins APL Technical Digest*, Vol. 34, No. 3, 2018, p. 14, https://dragonfly.jhuapl.edu/News-and-Resources/
4   The initial statement in the community announcement was "Up to three MMRTGs are available at the cost of $105M for one unit, $135M for two units, and $165M for three units," although these cost numbers were subsequently revised downward. Some proposers may have considered a design that used more than one MMRTG to be risky from a schedule perspective, because this would imply a rather higher production rate than had been exercised in the 2010s, and indeed the community announcement stated "the usage of

MMRTG(s) requires delaying the launch Readiness Date by at least one year to no earlier than 2025 to allow for mission-specific funding to support provision of MMRTGs."

5 Langelaan, J., Schmitz, S., Palacios, J., and Lorenz, R., "Energetics of Rotary-Wing Exploration of Titan," *IEEE Aerospace Conference*, Big Sky, MT, March 2017, doi:10.1109/AERO.2017.7943650.

6 Grubisic, A., Trainer, M. G., Li, X., Brinckerhoff, W. B, van Amerom, F. H., Danell, R. M., Costa, J. T., Castillo, M., Kaplan, D., and Zacny, K., 2021. Laser Desorption Mass Spectrometry at Saturn's moon Titan. *International Journal of Mass Spectrometry*, 470, p.116707.

7 Zacny, K., Lorenz, R., Rehnmark, F., Costa, T., Sparta, J., Sanigepalli, V., Mank, Z., Yen, B., Yu, D., Bailey, J., and Bergman, D., "Application of Pneumatics in Delivering Samples to Instruments on Planetary Missions," *2019 IEEE Aerospace Conference*, doi:10.1109/AERO.2019.8741887.

8 Lorenz, R. D., and Le Gall, A., "Schumann Resonance on Titan: A Critical Reassessment and Implications for Future Missions," *Icarus*, Vol. 351, 2020, p. 113942, https://doi.org/10.1016/j.icarus.2020.113942.

9 A useful overall source on this topic is Lee, Y., and Bairstow, B., *Radioisotope Power Systems Reference Book for Mission Designers and Planners*, Jet Propulsion Laboratory, National Aeronautics and Space Administration, Pasadena, CA, 2015, https://trs.jpl.nasa.gov/handle/2014/45467. However, its graphic on MMRTG performance appears overly pessimistic. See also Whiting, C. E., "Empirical Performance Analysis of MMRTG Power Production and Decay," *2020 IEEE Aerospace Conference*, doi: 10.1109/AERO47225.2020.9172270.

10 Lorenz, R., "Energy Cost of Acquiring and Transmitting Science Data on Deep Space Missions," *Journal of Spacecraft and Rockets*, Vol. 52, 2015, pp. 1691–1695. In fact, a cruder assessment of the energy per science bit metric, and its implications for the necessity of radioisotope power for Titan exploration, was a central theme of the paper in which I first laid out the Titan helicopter concept of operations: Lorenz, R. D., "Post-Cassini Exploration of Titan: Science Rationale and Mission Concepts," *Journal of the British Interplanetary Society*, Vol. 53, 2000, pp. 218–234.

11 Distance requirements on planetary missions tend to be capability-driven. See Lorenz, R. D., "How Far Is Far Enough? Distance Requirements Derivation for Planetary Mobility Systems," *Advances in Space Research*, Vol. 65, 2020, pp. 1383–1401.

12 Lorenz, R., "The Unnatural Selection of Planetary Missions," *The Space Review*, 2015 Aug. 17, https://www.thespacereview.com/article/2808/1

13 McGee, T. G., Artis, D. A., Cole, T. J., Eng, D. A., Reed, C. L., Hannan, M. R., Chavers, D. G., Kennedy, L. D., Moore, J. M., and Stemple, C. D., "Mighty Eagle: The Development and Flight Testing of an Autonomous Robotic Lander Test Bed," *Johns Hopkins APL Technical Digest*, Vol. 32, No. 3, 2013, pp. 619–635.

14 McGee, T. G., Adams, D., Hibbard, K., Turtle, E., Lorenz, R., Amzajerdian, F., and Langelaan, J., "Guidance, Navigation, and Control for Exploration of Titan with the Dragonfly Rotorcraft Lander," *2018 AIAA Guidance, Navigation, and Control Conference*, AIAA-2018-1330. See also ref. 14.

15 Lorenz, R. D., "Triboelectric Charging and Brownout Hazard Evaluation for a Planetary Rotorcraft," *AIAA Aviation Forum 2020* (virtual), AIAA-2020-2837. See also Rabinovitch, J., Lorenz, R., Slimko, E., and Wang, C., "Scaling Helicopter Brownout for Titan and Mars," *Aeolian Research*, Vol. 48, 2021, p. 100653.

16 Witte, I. R., Bekker, D .L., Chen, M. H., Criss, T. B., Jenkins, S. N., Mehta, N. L., Sawyer, C. A., Stipes, J. A., and Thomas, J. R., "No GPS? No Problem! Exploring the Dunes of Titan with Dragonfly Using Visual Odometry," *AIAA Scitech 2019 Forum*, AIAA-2019-1177.

17 On the general challenge of environment specification on planetary missions, see Lorenz, R., "How to Land on an Alien World," *Smithsonian Air and Space*, Vol. 36, No. 1, April/May 2021, pp. 22–27.

18 Lora, J. M., Tokano, T., d'Ollone, J. V., Lebonnois, S., and Lorenz, R. D., "A Model Intercomparison of Titan's Climate and Low-Latitude Environment," *Icarus*, Vol. 333, 2019, pp. 113–126.

19 Lorenz, R., "An Engineering Model of Titan Surface Winds for Dragonfly Landed Operations," *Advanced Space Research*, Vol. 67, pp. 2219–2230.

20 Lavely, A., Lorenz, R., and Schmitz, S., "Large-Eddy Simulation of Titan's Near-Surface Atmosphere: Convective Turbulence and Flow over Dunes with Application to Huygens and Dragonfly," *Icarus*, Vol. 357, 2021, p. 114229; Lorenz, R. D., Shiraishi, H., Panning, M., and Sotzen, K., "Wind and Surface Roughness Considerations for Seismic Instrumentation on a Relocatable Lander for Titan," *Planetary and Space Science*, Vol. 206, 2021, p. 105320. https://doi.org/10.1016/j.pss.2021.105320

21 See Lorenz, R. D., and Mitton, J., *Titan Unveiled*, Princeton University Press, Princeton, NJ, April 2008.

22 Lorenz, R., "Titan Surface Force Model for the Dragonfly Rotorcraft Lander," *Planetary and Space Science*, Vol. 214, 2022, 105449. https://doi.org/10.1016/j.pss.2022.105449

23 Corlies, P. M., McDonald, G. D., Hayes, A. G., Wray, J. J., Adamkovics, M., Malaska, M. J., Cable, M. L., Hofgartner, J. D., Horst, S., Liuzzo, L. R., Buffo, J. J., Lorenz, R. D., and Turtle, E. P., "Modeling Transmission Windows in Titan's Lower Troposphere: Implications for Infrared Spectrometers Aboard Future Aerial and Surface Missions," *Icarus*, Vol. 357, 2021, p. 114228.

24 Lorenz, R., and Lunine, J., "Titan's Snowline," *Icarus*, Vol. 158, 2002, pp. 557–559. The surface temperature at the equator is 93.7 K and varies by only $\sim$1 K. Although the freezing point of pure methane is $\sim$91 K, in the presence of 1.5 bar of nitrogen, the liquidus temperature of the mixture ($\sim$80 K) is reached only above about 14 km altitude. So although icing would be a concern for some Titan aviation, near-surface flights such as those of Dragonfly should not be affected.

25 Lorenz, R., Schmitz, S., and Kinzel, M., "Prediction of Aerodynamically-Triggered Condensation: Application to the Dragonfly Rotorcraft in Titan's Atmosphere," *Aerospace Science and Technology*, Vol. 114, p. 106738, https://doi.org/10.1016/j.ast.2021.106738

Chapter 8

DRAGONFLY DEVELOPMENT, SCIENCE, AND OPERATIONS

"Perfection is achieved, not when there is nothing more to add, but when there is nothing left to take away."

—Antoine de Saint-Exupéry (1900–1944), French aviator and author

Dragonfly will launch in 2027, encased in an aeroshell and attached in a cruise stage, much like Curiosity or Perseverance. Titan and Saturn are at a great distance from the Earth and the Sun [Saturn is at 9–10 astronomical units (AU), compared with 1.5 AU for Mars], and so it is impractical to launch on a trajectory directly there.[1] Accordingly, planetary gravity assists (or slingshot maneuvers) are required. In the current plan, permitted by a higher-capability launch vehicle than had been proposed originally, the spacecraft arcs out from Earth to beyond the orbit of Mars, where it makes a deep-space maneuver (a rocket burn of some hundreds of meters per second) using its own propulsion. This burn directs it back inward, where it picks up energy from a flyby of the Earth to then fly out to Titan (see Fig. 8.1).[*]

Unlike Cassini, Dragonfly does not go into orbit around Saturn (a maneuver that demanded considerable rocket propulsion capability). Rather, the mission exploits Titan's deep soft atmosphere to decelerate by entry directly from its interplanetary trajectory, like most Mars missions. One aspect of entry system design is that the loads (structural and thermal) are a strong function of entry velocity. Thus, it is important to minimize this quantity. An important way to do that for a Titan mission is to aim at Titan where its orbit around Saturn is receding from the spacecraft (Fig. 8.2), such that the vector difference of the spacecraft and Titan velocities is reduced.

---

[*]The trajectory is called a 3+ΔVEGA, an Earth gravity assist (EGA) with a deep-space maneuver (ΔV or *delta-vee*). The 3+ designation notes that the flyby is outbound and occurs 3 years after launch.

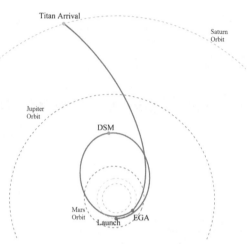

Fig. 8.1 With a high-energy launch in 2027, Dragonfly makes 1.5 revolutions around the sun with a deep space maneuver and an earth gravity assist flyby in 2030, reaching Titan by 2034. Reaching Titan requires about an order of magnitude longer flight in space than does a typical flight to Mars.
Source: Applied Physics Laboratory (APL).

Because Titan is tidally locked (like Earth's moon, with the same face always presented to its parent planet), this effectively limits a mission using "standard" (ie, Mars-like) heat shield materials and aeroshell designs to the trailing hemisphere (Fig. 8.3). Similarly, the entry cannot be too steep, although the properties of Titan's atmosphere are such that entry can be much steeper than at Mars. Thus, the available locus of target areas is ring-shaped. Inspection of the Titan map suggested an obvious target for Dragonfly—the Selk impact structure. Coincidentally, this is not too distant from the Huygens

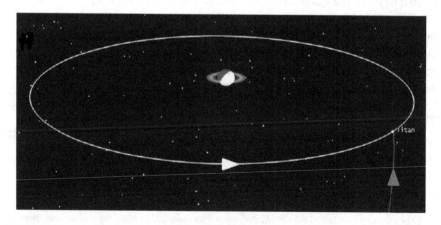

Fig. 8.2 The arrival in the Saturnian system is timed such that the spacecraft hits Titan's receding face, thus minimizing the entry velocity.
Source: APL.

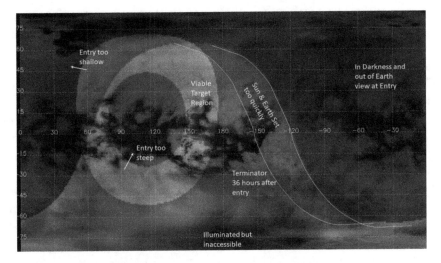

Fig. 8.3 Various factors define the region that can be targeted for Dragonfly's entry and descent: the annulus of acceptable entry angle intersects with the region in Earth view during and for a minimum period after entry. Zero longitude where Saturn is overhead is at the left and right edges of the map; the left half of the map is Titan's trailing hemisphere, the right is the leading hemisphere over which entry velocities would be much more challenging.
Source: Author, published in [3], CC BY-4.0.

probe landing site, and because Dragonfly will make its entry and descent (Fig. 8.4) at roughly the same Titan season as Huygens, this allows the Huygens data to strongly inform expectations of winds and atmospheric structure. Selk is on the anti-Saturn side of Titan, so although the landscape is expected to look spectacular, Saturn will unfortunately not be visible in the sky. (It would in any case be rather blurry, due to the optical depth of Titan's haze, which acts like a heavy overcast.)

Scientifically, the target area for Dragonfly (the Selk impact crater) was chosen because the cratering event is likely to have produced chemical compounds of prebiotic interest, where molten ice crust (ie, water) has interacted with surface organics.[2] Cassini data (Fig. 8.5) show that water-rich material is exposed around and in the crater.

However, although Dragonfly is equipped with hazard avoidance, aiming straight at the crater might put the vehicle over rather rugged terrain. Thus, it seemed prudent to aim the vehicle at a more benign area first, and then use its mobility to traverse to the target, armed with much better information than we have now from Cassini. Thus, the initial landing site[3] (Fig. 8.6) is planned to be among sand dunes about 100–150 km (60–100 miles) to the south-south-east of the crater.

The characteristics of Titan's dune fields,[4] known mainly from Cassini's radar instrument, are such that they are linear in form, resembling some of the

largest dunes found on Earth, those in the Namib desert in southern Africa and
the Arabian desert. The dunes are remarkably uniform in character and are
almost invariably oriented east–west (which actually reflects seasonally alter-
nating winds that are mostly south, then north, with a slight eastward bias).
They are separated by a rather constant 3–4 km (2–2.5 miles), likely reflective
of the planetary boundary layer structure (see Chapter 2).

The tallest dunes known on Titan, covering about 15% of Titan's surface in
a broad equatorial belt (Fig. 8.7), are about 150 m high, although the morphology of the dunes near Selk, which sort of taper out (Fig. 8.8), suggests that
they will be rather shallower than this. In any case, digital elevation models of
the Namib (which has dunes up to 150 m high) were used as analogs of Titan
terrain for simulating Dragonfly's optical navigation system. Those data show
that at a 30-m scale at least, some 50% of the area has a slope of 1 deg or less,

**Fig. 8.4 Entry descent and landing sequence. The cruise stage separates some tens of
minutes prior to entry. After the vehicle passes peak load and peak heating, a small
drogue parachute is deployed to stabilize the vehicle over the transonic speed regime.
Dragonfly descends under the drogue, still in its aeroshell, over the next 1.5 h or so before
deploying a large parachute to slow its descent at about 4 km altitude and allow the heat
shield to safely fall away. After preparing the rotors/motors for operation and acquiring
sensor data to judge ground-relative position and motion, the vehicle drops from the
backshell to make the transition to powered flight, and lands under rotor power.**
Source: APL.

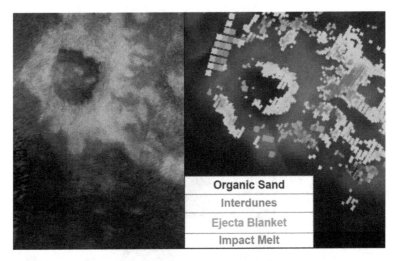

Fig. 8.5 The Selk impact crater, 80 km across, seen in a false-color near-infrared composite of Cassini data from the visual and infrared mapping spectrometer (VIMS), with red, green, and blue representing light at 5, 2, and 1.3 μm, respectively. The blue tints represent water-ice-rich material of particular astrobiological interest.
Source: APL, courtesy Shannon MacKenzie.

and 95% is less than 6 deg. For a vehicle able to tolerate modest slopes (e.g., 10 deg), there are certain to be ample locations that permit safe landing.

Once safe landing on arrival is achieved, the rotorcraft mobility capabilities can be exercised progressively—for example, first making a brief hop for a few seconds within the immediate vicinity of the landing site where the terrain will be known from panoramic and/or descent imaging. Depending on the heterogeneity of the surface (e.g., patches of sand), a small displacement of a few meters or tens of meters may enable the sampling of different materials.

Then, flights of progressively increasing duration, range, and/or height can be made, returning to the original, known-safe landing site. These flights can assess the performance of various sensors; for example, an initial hop may be made using inertial guidance alone, whereas later flights use optical navigation only after the quality of in-flight imaging and the abundance of suitable landmarks on Titan have been verified.

If the Titan terrain is as benign as the Namib analog suggests, safe landing zones can be more or less guaranteed between the dunes, and the flight range of the vehicle can be exploited using the following strategy (Fig. 8.9):

1. A second landing zone (CS1) is identified by ground analysis of reconnaissance imaging, a distance R/3 or less away from the initial landing site (LS1).
2. The vehicle makes a sortie over this zone using its sensors (lidar for terrain roughness, imaging, etc.) and returns to the original landing site LS1.

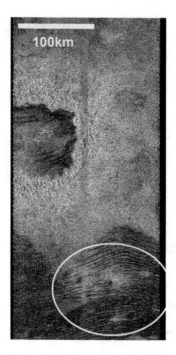

Fig. 8.6 Dragonfly's initial landing site will be somewhere in the ellipse at lower right, overlain on a Cassini radar image. The dark streaks are dunes of organic sand, roughly 3–4 km apart. Dragonfly will explore the dunes, and then the impact ejecta deposits, as it traverses north-northwest toward the center of the 80-km Selk crater. The terrain becomes very rugged at the crater rim, but the Dragonfly concept of operations allows terrain to be scouted for safe landing sites before committing to a new area. Depending on where in the ellipse the first landing occurs, the traverse to the crater may require more than 150 km of travel.
Source: Author.

3. Analysis on the ground of the sensor data confirms one or more safe sites within zone CS1 (or if no satisfactory site is found, return to Step 1).
4. A candidate third landing zone (CS2) is identified in reconnaissance imaging, a distance 2R/3 away from LS1.
5. The vehicle makes a sensing sortie over CS2 but lands at CS1, which is now LS2.

In this way, the mission need not commit to landing sites that have not first been assessed to be safe. This conservative approach, although taking longer to achieve a given multihop traverse range, enables the contemplation of much rougher terrains that may be associated with more appealing scientific targets (eg, cryovolcanic features or impact melt sheets where liquid water may have interacted with organics on Titan).

At each new landing site, the high-gain antenna (HGA) is unstowed and downlink begins. Priority data might include flight performance information

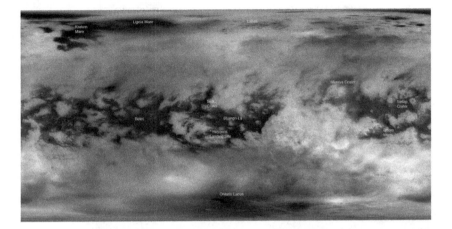

Fig. 8.7 Titan map made from Cassini near-infrared images. The map (cylindrical projection, so high-latitude features are considerably stretched) is centered on the anti-Saturn point, with zero longitude at the left and right edges. The delivery circumstances of both Huygens and Dragonfly favor delivery near the center, at the edge of the Shangri-La sand sea. The bright region Xanadu is on Titan's leading hemisphere and was first detected in ground-based telescopic observations, then imaged by the Hubble space telescope in the early/mid-1990s.
Source: Author.

and aerial imaging of the landing site to confirm its exact location on maps made from prior reconnaissance. A quick-look site assessment would use sensors on the landing skids to estimate the surface properties (see later this chapter). These measurements would take only seconds to minutes. Over a period of a few hours, the neutron-activated gamma-ray spectrometer would determine the bulk elemental composition of the landing site, allowing identification among a number of basic expected surface types (e.g., organic dune sand, solid water ice, frozen ammonia-hydrate, etc.).

Armed with this information and imaging to characterize the geological setting, the science team on the ground might elect to acquire a surface sample with one of the drills and analyze it with the mass spectrometer. This is a relatively energy-intensive activity, because the heavy drill motor must be warmed up for operation, and the pneumatic blowers consume several hundred watts to ensure a vigorous flow through the sample transfer hose.

The DragonCam suite includes down-looking wide-angle cameras on the vehicle's belly to view the area under the lander to decide on sampling and possible seismometer deployment. A close-up imager looks at the sites where each drill will touch the ground and is able to resolve individual sand grains. Night-time imaging is performed using light-emitting diode (LED) illuminators like those flown on Phoenix and Curiosity. These would permit better color discrimination of Titan surface materials (because the daytime illumination, filtered by the thick atmospheric haze, is predominantly of red

Fig. 8.8 Top: Spaceborne radar image (X-SAR) of the Quattaniya Dunes, west of Cairo, Egypt. These show the same linear morphology as those on Titan, as well as the tapering shape indicative of a limited sand supply that is observed in the dunes south of Selk. Thus, the Titan dunes probably have a similar size and morphology to those seen in the middle and lower panels, which show kiteborne and ground-level views, respectively. Although the dune ridge has steep slopes, there are wide, sandy, shallow-sloped plinths that grade into the interdune plains.[†]
Source: Author [3] CC BY-4.0.

light) and could use ultraviolet illumination to help identify surface organic material via fluorescence, which is common in the polycyclic aromatic hydrocarbons (PAHs) expected in the dune sands.

The DraGMet geophysics and meteorology package will monitor the weather. It has sensors to measure the abundance of hydrogen and methane (the latter being analogous to measuring the humidity on Earth). It also records pressure and temperature, as well as winds. The latter measurement is per-

---

[†]It was near these dunes that Antoine de Saint-Exupery crashed his plane in 1935 during an attempt to beat the speed record of the Paris-to-Saigon air race. He and his navigator nearly died of dehydration before being found by a Bedouin after 4 days. This experience inspired the opening desert scene of his children's book The Little Prince, which features an asteroid.

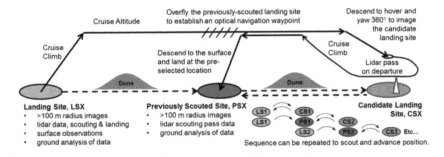

Fig. 8.9 Leapfrog reconnaissance and survey strategy enable potential landing sites to be fully validated with sensor data and ground analysis before being committed to.
Source: APL.

formed by hot film anemometry, basically heating a small cylinder and measuring how much wind chill the cylinder feels in different directions. Because the lander structure can distort the airflow slightly, DraGMet has four wind sensors, one on each rotor hub, so that at least two are upwind of the vehicle and see undisturbed airflow. Similar wind sensors are used by the lander to check that conditions are safe for takeoff.

Sensors on the skids measure some basic properties of the ground (dielectric constant and thermal properties), in part to diagnose whether the material is damp, which might influence our decision of whether to acquire a sample. Additionally there are geophones (rugged but relatively insensitive seismometers). If the ground is sufficiently flat beneath the lander, and it is expected that the lander will remain for a prolonged period, the team may elect to deploy the seismometer onto the ground. The seismometer, a contribution from the Japanese Aerospace Exploration Agency (JAXA), is more than a hundred times more sensitive than the geophones and so will have a much better chance of detecting Titanquakes and perhaps measuring the ice crust thickness. In order to realize this sensitivity, however, it needs to be decoupled from the lander and the vibrations of its thermal control systems (and movement of the lander induced by the wind). Thus, the seismometer is lowered on a flexible tether by a small winch on the lander belly. As with the DrACO drills, the accommodation of a seismometer brings challenges that are relatively unfamiliar to rotorcraft and spacecraft designers alike!

Note that although Dragonfly lacks a robotic arm, it can nonetheless manipulate surface materials to understand their physical character. One example is that the geophones and seismometer can observe the noise transmitted through the ground during drilling, diagnostic of the mechanical properties of the regolith and possibly indicating near-surface layering. Another example is that one or more rotors can be spun (at progressively higher speeds) to induce a known downwash on the surface material, and the speed at which sand grains begin to move (indicated either by imaging or a skid-mounted photodiode that will record the shadows of the blowing grains) can thereby be determined. This *saltation threshold* is a key parameter in interpreting the large-scale morphology and orientation of Titan's dunes in global circulation models. There are indications that, as on Earth, because large dunes take tens of thousands of years to form or reorient, the dune pattern carries some memory of past climate; models suggest that astronomical changes (Croll-Milankovitch cycles, similar to those that drive the ice ages on Earth) may alter Titan's wind patterns and indeed the geographical distribution of its surface liquids. Decoding the dune pattern, however, requires good knowledge of the saltation threshold (estimated to be around 1 m/s, but laboratory measurements on Earth are limited in their capability to replicate Titan conditions, to say nothing of our ignorance of the exact sand composition and the possible role of triboelectric charging).

Overall, Dragonfly will spend less than 1% of its time actually flying. Most of the time the lander is in a hibernation state, recharging its battery (although the DraGMet sensors are always running), but there are a great many scientific observations to plan. The observations, and their energy demands, are tracked in a planning tool (Fig. 8.10). This also estimates how much data each activity will produce and ensures that the communications sessions with ground antennas can keep up with the rate of data generation. These observation plans will surely change once we arrive at Titan and learn new things, but example scenarios must be used even in early development to identify the required capabilities.

In the nominal plan, Dragonfly will fly every other Titan day (Tsol), that is, about once per month. It is expected that the forward flight range that can be achieved is several kilometers, even taking the conservative two-steps-forward-one-step-back leapfrog approach. Thus, during its more than three-year mission, Dragonfly is expected to perform about 40 flights, adding up to 150–200 km of flights from the landing ellipse, perhaps to the interior of the Selk crater. During the Tsols between flights, Dragonfly spends more time downlinking data (because there is more energy available to spend on downlink) and conducting measurements.

Perseverance and other Mars rover missions can be grueling on their operators, requiring rapid turnaround of scientific planning and (in their early stages at least) working on "Mars time" (which given that a Mars day is about 40 min longer than a terrestrial day, leads to a perpetual jet lag). However, for Dragonfly there will typically be a busy week where commands will need to be generated between daily Deep Space Network passes, and then there will be a quiet week during the Titan night, when no communication is possible. This time will, of course, be used in part for analyzing data and planning future operations, but is a much more comfortable operating cadence that can be sustained for years.

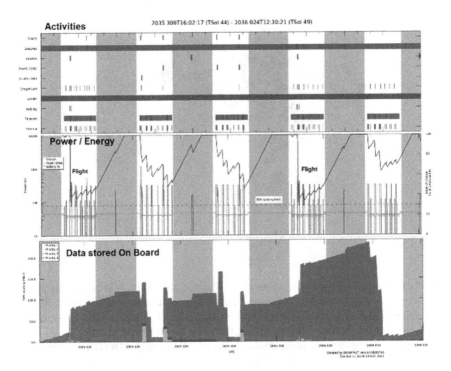

Fig. 8.10 An example timeline of five Titan days (Tsols) during the mission. A planning tool (DraMPACT) is used to track the vital resource (energy) and the product (science data) to ensure the books are balanced. In this plan, flights take place on Tsols 44 and 48 during the day (white background). The power draw (middle panel) of flights results in a rapid partial depletion of the battery (vertical drops in blue line), which is recharged overnight (gray bands). The flights and science observations (bars in upper panel) produce data that is held on a solid-state recorder (red bar chart in bottom panel) until it is transmitted to Earth during downlinks (which produce most of the sawteeth in the blue line in the middle panel). In this five-Tsol plan, which ensures the battery is never too deeply depleted (30% minimum state of charge), all the data acquired on the ground, including the energy- and data-heavy GCMS (Gas Chromatograph Mass Spectrometer) analysis on Tsol 45 and in the flights are returned by the end of Tsol 49.
Source: APL, courtesy Scott Murchie.

There is no "sudden death" at the end of the nominal mission, and certainly there is ample precedent of planetary missions being extended for years. The multimission radioisotope thermoelectric generator (MMRTG) output will decline, ultimately to the point where there is no longer enough heat to maintain components in their operating ranges, but that would likely be several years beyond the nominal. Of course, mechanical components may progressively wear out, and Dragonfly might (like the Spirit rover) lose the ability to relocate long before it loses function altogether. In that sense it would become a fixed lander, but still with much scientific capability.

Fig. 8.11 Dragonfly's overall configuration in Sept. 2021 (still a year before the preliminary design review, so further changes will likely occur). The nose of the vehicle has been streamlined, and a medium-gain antenna looking somewhat like a paint roller is at the back. Bracing struts have been added to stiffen the landing gear, and the dark red pipe down the side of the vehicle is the duct to bring heat from the MMRTG forwards.
Source: APL.

DESIGN EVOLUTION

Upon selection in June 2019, NASA directed the Dragonfly team to plan for launch in 2026 [rather than the 2025 launch date stipulated in the announcement of opportunity (AO)]. It was determined that a larger (5-m) fairing on the launch vehicle would be provided, allowing the aeroshell to grow and to accommodate a slightly longer lander body (Fig. 8.11). This would provide more pitch and yaw control authority from the rotors and would allow a more streamlined shape. The greater aeroshell volume also allowed the lander legs to be fixed, avoiding the hinge mechanism and the need for the DrACO tubing to accommodate that movement.

As delays and additional costs due to the COVID-19 pandemic accrued across NASA's mission portfolio in 2020 and 2021, in summer 2021 NASA directed the team to plan for launch in 2027. This rephasing of the project's development deferred NASA's peak spending on Dragonfly to relieve the pressure on its immediate budget, but also gave the project more time to spend on the technology maturation and detailed design phases. NASA committed a (to-be-determined) heavy launch vehicle with enough capability to send Dragonfly to Titan without requiring a Venus flyby. This, in turn, permitted a shorter cruise duration such that an arrival at Titan by 2034 would be maintained, despite the later launch. The thermal design of the cruise stage and aeroshell would also be simplified, because the spacecraft would not have to

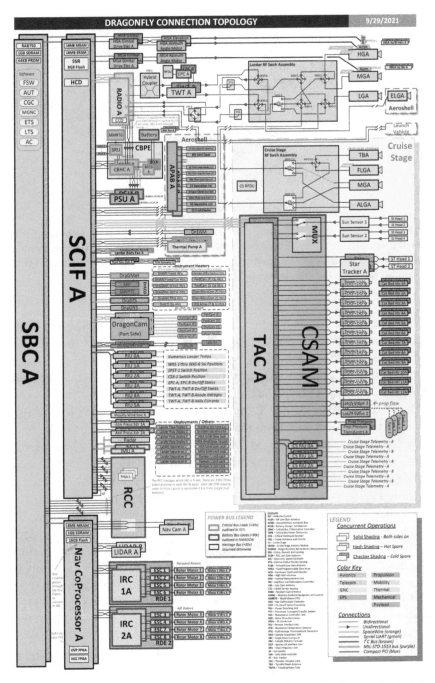

Fig. 8.12 Even with its details barely visible on the printed page, the Dragonfly electrical connection topology diagram makes clear the complexity of the vehicle and its interface with the cruise stage. Double boxes such as the single-board computer (SBC) indicate redundant units for enhanced reliability.
Source: APL

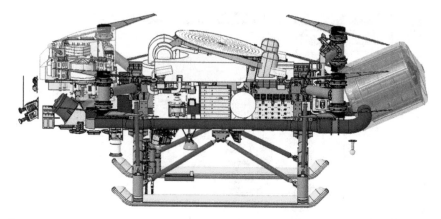

Fig. 8.13 Side view of the Dragonfly computer-aided design (CAD) model, with the hull rendered transparent. The formidable number of components and systems crammed into the body is evident. In this design iteration, the duct pulling heat from the MMRTG at right is evident as the brown horizontal pipe.
Source: APL.

fly as close to the Sun as it would in the original mission profile. The shorter cruise duration would mean the mission would land with a fresher MMRTG (the output of which declines by ~2% per year, in part due to decay of the ^{238}Pu fuel and in part due to degradation of the thermoelectric energy converters inside the unit), a very valuable outcome.

Detailed analysis of the heat transfers led designers to lean toward using Titan air to transport heat from the MMRTG to the equipment in the lander body, rather than a pumped liquid (Fig. 8.13). This sort of thermal management is, of course, used all the time in computer equipment. In space applications, it was used by the Soviet radioisotope-heated Lunokhod-1 and -2 lunar rovers in the early 1970s. Those had sealed hulls, with internal air driven by a fan.

Another change from the Phase A design was the introduction of a medium gain antenna. This provided a backup communications pathway in case of problems with the HGA. Inevitably, as detailed designs were developed (Fig. 8.12), the projected mass of Dragonfly increased and will probably end up only a bit less than the ~900 kg Curiosity rover. Such mass growth is typical for space projects at this stage. At the time of writing, Sept. 2021, the project is still a year away from its preliminary design review (PDR) when most details can start to be considered to be firm.

The programmatic aspects of Dragonfly also begin to fall into place. For example, while the AO solicited proposals using an MMRTG, and the physics make it the obvious choice, the launch of nuclear material into space is a serious matter, and must be approved at the highest levels of government per the National Environmental Policy Act (NEPA). Similarly, while Titan is far too cold for any terrestrial biota to thrive, the U.S. is a signatory of the 1967 Outer Space Treaty, and any possible protection of extraterrestrial habitable

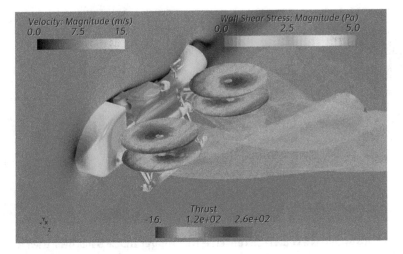

Fig. 8.14 Computational fluid dynamics (CFD) simulations are being used to evaluate streamlining options to minimize drag and thus maximize the range that can be accomplished in flight. One complicating factor in the octocopter design is that the lower rotors are immersed in the downwash of the upper ones, and this wake impingement depends on the attitude and forward speed. Notice the azimuthal asymmetry in the thrust distribution (numbers are blade loading in Newtons per meter) on each rotor—the inboard/outboard position of the thrust maximum on each pair is mirrored because the upper/lower rotors rotate in opposite directions; therefore, the side of the advancing rotor is different for each.
Source: Image courtesy Wayne Farrell and Mike Kinzel, University of Central Florida.

environments must be considered. This falls under the purview of the NASA Planetary Protection Officer, who must make a determination of the mission and any necessary measures such as analysis or sterilization. Finally, international agreements must be drawn up to formally secure foreign contributions to Dragonfly. These include the JAXA Seismometer, a package of instrumentation to measure the hypersonic aerothermodynamic environment of the heatshield from the German Aerospace Agency DLR, and contributions of gas chromatography hardware for DraMS from the French Space Agency CNES. As Wernher von Braun once said "We can lick gravity but sometimes the paperwork is overwhelming".

Meanwhile, hundreds of people are working on the many complex parts of Dragonfly, resolving the details of the rotors and motors, the formidable electrical system needed to drive them, the thermal management of the vehicle, and so on.[‡] Computer simulation is, of course, a significant component of these and other analyses (e.g., Figs 8.14–8.16), but a lot of testing is also underway in facilities at APL and elsewhere (Figs. 8.17–8.18).

[‡]For example, pushing ~10 kW of power out to the motors for flight means, even with a high-voltage (~100 V) bus, many tens of amps of current, necessitating a lot of copper. But this good electrical conductor must not also allow too much heat to leak out of the vehicle. Thus, even something as superficially simple as a wire needs careful design attention to its routing, insulation, and so on.

Fig. 8.15 Brownout is a concern, as on terrestrial rotorcraft. CFD has been used to quantify the shear stress on the surface and track lofted material to assess the amount of erosion that might occur under the lander and the amount of dust that might be deposited on lander surfaces (especially camera windows). Interestingly, the impingement of the four separate downwash tubes from the four sets of rotors leads to a convergent flow under the lander and a fountain of dust that impinges on its lower surface.
Source: Image courtesy Mike Kinzel, University of Central Florida.

One of the virtues of a rotorcraft vehicle is that it is much easier to test on Earth than is a rocket-powered lander. Thus, scale model vehicles, using actual camera or navigation systems, can be repeatedly rehearsed in the field to flush out software bugs and failure modes. The Dragonfly team uses a half-scale vehicle (Figs 8.19–8.20) for much of its testing, which is flown at a facility near APL, but also occasionally at field locations.

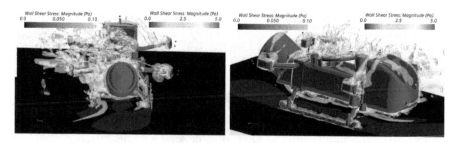

Fig. 8.16 Dragonfly is to investigate Titan's meteorology, and CFD simulations are used to predict the perturbation to wind sensors (red dots) mounted in different candidate locations. By having four DraGMet wind sensors outboard of each rotor support, one and usually two sensors are upwind of the lander structure and thus clear of the worst flow disturbances (shown as white clouds).
Source: Graphic courtesy Mike Kinzel, UCF.

Fig. 8.17 CFD predictions for the rotor performance, and in particular the thrust performance in cross-flow, where the downwash from one rotor is advected across the lower rotor, must be validated with wind tunnel tests. The COVID-19 pandemic brought considerable complications to such efforts, with access to NASA facilities restricted to special cases. Here, rotor tests were performed in the 14×22 ft wind tunnel at NASA Langley Research Center.
Source: APL.

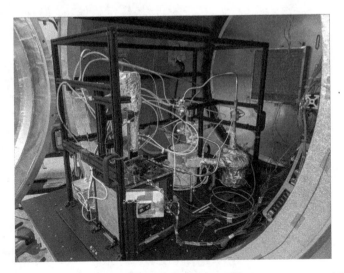

Fig. 8.18 The Titan pressure environment chamber (TPEC) was set up at APL to perform hardware tests not only at low temperature, but also at Titan pressure, to replicate more faithfully the gas and heat transport conditions. Early hardware tests included thermal management systems, rotor motors, landing dampers, and here, the DrACO sample transfer system.
Source: APL.

Fig. 8.19 ITP (Integrated Test Platform) drone—a half-scale Dragonfly—in tests at the Imperial Dunes in Southern California. The principal goal of these flight tests was to obtain navigation camera data over representative terrain. The black drone box in the center of the vehicle houses cameras and other sensors. Note the Federal Aviation Administration (FAA) registration number N886UH—this test vehicle weighing over 50 lb required a fully trained pilot- in-command for safety.
Source: APL.

Fig. 8.20 ITP drone in flight near Yuma, Arizona. A few seconds of flying takes the drone beyond the dune-buggy tracks (which act as artificially effective reference points on the otherwise rather bland dune terrain). A great advantage of a rotorcraft vehicle is the extensive amount of flight testing that can be performed compared with a rocket-powered lander.
Source: APL.

REFERENCES

1   The design of the original trajectory in the Step 1 proposal and Phase A study is described in Scott, C. J., Ozimek, M. T., Adams, D. S., Lorenz, R. D., Bhaskaran, S., Ionasescu, R., Jesick, M., and Laipert, F. E., "Preliminary Interplanetary Mission Design and Navigation for the Dragonfly New Frontiers Mission Concept," AAS 18-416, AAS Astrodynamics Specialist Conference, Snowbird, UT, 19–23, Aug. 2018.

2   Neish, C. D., Lorenz, R. D., Turtle, E. P., Barnes, J. W., Trainer, M., Stiles, B., Kirk, R., Hibbitts, C. A., and Malaska, M. J., "Strategies for Detecting Biological Molecules on Titan," *Astrobiology*, Vol. 18, No. 5, 2017, pp. 571–585. doi:10.1089/ast.2017.1758

3   Lorenz, R. D., MacKenzie, S. M., Neish, C. D., Le Gall, A., Turtle, E. P., Barnes, J. W., Trainer, M. G., Werynski, A., Hedgepeth, J., and Karkoschka, E., "Selection and Characteristics of the Dragonfly Landing Site near Selk Crater, Titan," *Planetary Science Journal*, Vol. 2, 2021, p. 24. https://doi.org/10.3847/PSJ/abd08f

4   All you could ever want to know about sand dunes is in Lorenz, R. D., and Zimbelman, J., *Dune Worlds: How Wind-Blown Sand Shapes Planetary Landscapes*, Praxis-Springer, May 2014. All you could ever want to know about Titan is in my Saturn's Moon Titan: Owners Workshop Manual, Haynes, Yeovil, UK (2020).

Chapter 9

THE OUTLOOK FOR PLANETARY AERONAUTICS

"Once you have tasted flight, you will forever walk the earth with your eyes turned skyward. For there you have been, and there you will always long to return."

—Leonardo da Vinci (1452–1519)

We have reviewed the specifics of the two planetary rotorcraft vehicles in development or operation, namely Dragonfly and Ingenuity. Predictably, much has been said of a "new era" of aerial planetary exploration, heralded by these two missions. As the reader has learned, the two vehicles have very different objectives, design constraints, and flight environments—indeed, one is 400 times more massive than the other! Before we consider the future possibilities on Mars and Titan, let us first consider other planetary destinations with atmospheres, namely Venus and the outer planets. It should be underscored that what follows is not the determination from some august panel of experts, but my opinion only.[1]

OTHER DESTINATIONS: VENUS

One parameter of the Venus atmosphere is attractive from an aeronautical perspective—the density. At the surface, it is about 12 times denser than Titan's atmosphere, or 50 times denser than Earth's. This density results in part from the high molecular weight (CO_2 being 44, compared with nitrogen's 28) but mostly from the crushing pressure of 90 bar, equivalent to that on the sea floor on Earth at a depth of 900 m (3000 ft).

The pressure is not, per se, a problem for the spacecraft designer. High pressures are routinely confronted by deep sea systems—some attention must be paid to dimensional changes (ie, if parts are squeezed by high pressure, electrical parameters such as capacitance or resistance may change; any mechanical or electromechanical systems subject to differential contractions might seize up), but otherwise the pressure itself is not a threat. Rather, excluding conductive salt water (in the case of our seas) or hot, corrosive gas (on Venus) leads a typical designer to build the vehicle hull as a sealed pressure vessel. There may be other ways, with non-load-bearing hulls, to exclude the external threat; for example, some deep-sea systems simply fill the equipment cavity with oil.

It is, however, the temperature on Venus that is the problem. At some 750 K, it is above the melting point of lead! With a thick, hot atmosphere in all directions, a vehicle near the Venus surface has nowhere to reject the heat generated by its systems or that leaks in from the environment, unless it can actively pump heat to a "radiator" that is hotter than the Venus ambient temperature. Such active refrigeration would be energy-intensive and require very high-temperature materials and working fluids—essentially the type of equipment used in nuclear reactors. Although some theoretical designs of such systems have been sketched out,[2] they would require formidable investment to realize.

Thus, Venus probes to date have simply relied on the thermal transient—of having enough thermal mass and enough insulation that they warm up toward Venus ambient temperatures only slowly. The roughly 1-t Soviet Venera and VEGA landers of the 1970s and 1980s lasted 1–2 h before their internal temperatures rose to the point at which electronics ceased operating (typically 80–200°C).

There have been some recent developments in high-temperature electronics that might allow some limited functionality at Venus temperatures (e.g., a low data rate from a weather station or simple seismometer modulated directly onto a radio link, perhaps powered by a wind turbine or high-temperature solar panel). The sodium-sulfur battery chemistry is one that can function at Venus temperatures, so conceivably the Ingenuity/Dragonfly concept of operations of trickle-charging a battery could support the energy needed for flight. But sophisticated processing, and certainly anything like the sensing functions needed to execute autonomous flight, seem well out of reach for the foreseeable future.

Up near the cloud tops on Venus, 60 km (200,000 ft) above the surface (Fig. 9.1), conditions are not too different from Earth sea level, and sustained operations without heroic thermal management are possible (as demonstrated by the VEGA balloons).[3] Being above the thick clouds makes solar power feasible, and there have been several studies of solar-powered Venus airplanes.[4] Provided a platform can spend enough of its time above most of the clouds, solar power is possible, although the atmosphere at these altitudes rotates about once per four Earth days, so at low latitudes at least, the platform will be in darkness for about 48 h out of 96. This places severe demands on energy storage for a heavier-than-air platform to sustain operations during that time, although it may be tolerable to glide down to a lower altitude rather than require level flight throughout.

Fig. 9.1 Artist's impression of a solar-powered aircraft soaring just above the main cloud deck in the Venus atmosphere, where unlike the torrid depths, temperatures are relatively benign.
Source: Les Bossinas, InDyne, Inc., 2001.

In any case, being so far from the surface, the advantages of a rotorcraft are not obvious. Further, the high molecular weight of the CO_2 atmosphere means that light-gas (hydrogen or helium) balloon solutions can provide sustained buoyant lift, limited only by leakage through the balloon envelope (perhaps Teflon-coated to avoid corrosion effects, with a metal film layer to inhibit diffusion). Thus, depending on the scientific objectives, a lighter-than-air solution may be the best approach to Venus aviation. Hybrid variants have even been proposed—vehicles with inflatable wings.

OTHER DESTINATIONS: OUTER PLANETS

There are altitude ranges in the giant planets Jupiter and Saturn, and the ice giants Uranus and Neptune, where the pressure and temperature conditions may allow reasonable heavier-than-air flight. One point that does deserve mention is that these outer planets have atmospheres that are predominantly hydrogen and helium (i.e., gases with a very low molecular weight compared with nitrogen and carbon dioxide). Thus, a balloon filled with pure helium brought in tanks from Earth would be negatively buoyant! Further, the gas on these worlds at a given pressure and temperature has a density that is ∼14–22 times lower than the atmospheres of the terrestrial planets at those conditions. It follows that to contain enough air to achieve some amount of buoyant lift by heating the atmosphere (ie, a hot air balloon or Montgolfière), a balloon envelope must be very large. Thus, although it is

Fig. 9.2 The small propeller on the nose of the World War II Messerschmidt Me-163 Komet fighter is a RAT to generate electrical power from the energy in the slipstream. Such a measure was needed because the vehicle used a short-duration rocket engine for climb and combat, and had to glide back to base.
Source: U.S. Air Force.

not completely impossible to construct a balloon platform for the outer planets,* it is demanding indeed.

A limited-duration heavier-than-air platform could, of course, be imagined; however, because these worlds lack true surfaces, the advantage of a rotorcraft over a fixed-wing solution is not obvious. Thus, the Mars airplane idea in Chapter 1 could be reimagined for Jupiter, although the questions of scientific value, range, lifetime, and data return would need close examination.

It is conceivable that a continuously flying airplane with a radioisotope power source could work, much like the AVIATR concept once considered briefly for Titan; however, the outer planet gravities are similar to those of Earth, so the required lift is higher than needed for Titan. The low wing loading required to bring the flight power low enough for a practical power source might require large wings that need complicated deployment by folding, internal inflation, or external inflation (like a parawing). Such aircraft might make an interesting design study, but are beyond the scope of this book.

One concept that has been considered for the outer planets, with a vague connection to rotorcraft, is the notion that deep atmospheric probes might use a ram-air turbine (RAT) for electrical power. Terrestrial aircraft use such devices—basically small fans exposed to the airflow of flight—in a few

---

*A famous study of an interstellar probe, Daedalus, by the British Interplanetary Society in the 1970s imagined that a nuclear hot-air balloon floating in the Jovian atmosphere would be used to harvest helium-3 for the fusion propulsion needed to accelerate the probe to near-relativistic speeds

circumstances (see Fig. 9.2). These include self-contained systems like defensive jamming pods, where it may be convenient to allow installation on a variety of platforms and not require them to provide power, but rather have the power generated by a RAT on the pod itself. Another circumstance is as an emergency backup in case of engine failure; because a RAT generates significant drag, it is usually recessed within the airplane hull and only extended into the airflow in the event of a contingency. The RAT provides power to maintain communication and control systems, and perhaps to aid in engine restart attempts. A RAT was also used on the Me-163 rocket fighter, because there was no conventional rotating or reciprocating engine to drive a generator.

Such a system on a bomb-shaped planetary probe is effectively a high-wing-loading rotorcraft descending in a windmill regime. In steady-state, the probe's weight is balanced by the drag (lift) on the rotor disk. Obviously, the bigger the disk, the slower the probe will descend and thus the less energetic will be the airstream. The advantages of such a system over stored energy like a battery are not obvious, because some sort of battery will likely be needed for predescent operations (like the deployment of the rotor) anyway, and a RAT system introduces complications for testing.

MARS

Ingenuity was explicitly a technology demonstrator. Now that that card has been played, any future rotorcraft at Mars must justify itself on utilitarian grounds. A natural step is to explore how the range/endurance of Ingenuity might be improved upon, and in particular how much of an instrument payload or sample acquisition system it might be able to carry.

A Mars Science Helicopter study[5] by the Jet Propulsion Laboratory (JPL), NASA Ames Research Center, and AeroVironment suggested one possibility to be a ~30-kg (66 lb) hexacopter (see Fig. 9.3). A 2 ft (0.64 m) 0.64-m rotor radius was considered at the Jezero Crater in the Martian northern hemisphere spring (0.015 kg/m^3, 220 K). The power needed is some 6.2 kW. At this size,

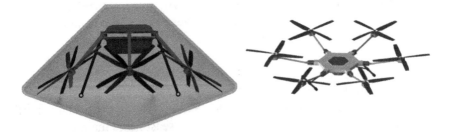

Fig. 9.3 Among future Mars Science Helicopter concepts being explored was a hexacopter configuration that would fold its rotor arms to maximize the rotor disk area that could be accommodated inside a heatshield.
Source: NASA.

the design can trade payload for battery weight (total energy), and trade range for hover time in the mission; thus, it could carry a 5-kg instrument payload for 10 km (or 4.7-min hover), or instead carry 2 kg for 16 km (10 miles, or 7.8-min hover)—in theory, at least.

Although the Mars Science Helicopter study considered aeroelastic aspects of the rotors, the report is silent on the question of thermal management. It seems (to this author at least) that the power dissipation in the propulsion motors is a fundamental limitation to aircraft operations on Mars, and to rotor-craft in particular. In so far as motor mass may be proportional to power, the heat capacity of a given mix of copper and structural metals such as beryllium or aluminum will also be proportional; thus, all else being equal with no loss of heat to the environment, an upscaled motor would have as many seconds to operate before it reaches unacceptable temperatures as does the Ingenuity motor (at the time of writing, a limit of about 200 s). Thus, the flight durations of 10 to 20 min envisaged for reduced-payload versions of the Mars Science Helicopter designs are likely to be difficult to attain.[†]

It is possible, especially at high temperatures, that some useful amount of heat can be rejected to the environment despite the thin atmosphere. There is a design contradiction between features (filters or seals) to exclude dust from the motor and encouraging vigorous airflow through it, but perhaps some arrangement of ducts and/or radiator plates can provide some cooling. However, the basic geometric consideration of mass:area means things get more and more difficult as you attempt to scale up. The only way to break out of the problem is to introduce some sort of liquid cooling system, a complication that adds mass, complexity, and cost. Another approach might be simply to take "breathers"—to perform as long a traverse as is permitted by the battery but in short hops, between which the vehicle cools down for 30 min or so. However, this increases the mission complexity and the number of takeoffs and landings, with associated implications for risk and reliability.

Note that as a vehicle gets larger, the Ingenuity approach of deploying it from a lander or rover becomes impractical. The larger dimensions of the vehicle become impossible to accommodate within the geometric envelope available, and the mechanisms to secure the vehicle for launch and to deploy it onto the ground become prohibitively heavy. Thus, the anticipated scenario for such a future vehicle would be in-air deployment[6] before the delivery system reached the ground—in short, the approach used by Dragonfly.

Some regions on Mars are more challenging than others. In particular, the southern hemisphere is at higher elevation and is more rugged than the north. The typical atmospheric density (depending on season and the specific eleva-

[†]Ben Pipenberg explained to me that making the Ingenuity motors a little longer would both increase their efficiency (and thus lower the amount of heat dissipation) while increasing their heat capacity. The combined effects would increase the permissible operating time by an appreciable factor, at the expense of a modest mass increase.

tion) in the southern highlands is 0.01 kg/m^3, compared with 0.015 kg/m^3 in the northern lowland plains. Inspection of the actuator disk equation (see Appendix) shows that this means either an increase in rotor diameter or an increase in shaft power of ~25% to compensate for the thinner air. (It should be noted that the Martian polar layered terrain, one of the most interesting targets for future exploration, is at high elevations.)

A recent NASA Ames study[7] examined possible rotorcraft designs to explore the highlands. The Mars Highlands helicopter would have 2-m-diameter coaxial rotors, these being the largest that could be accommodated inside a Mars Pathfinder aeroshell without folding or telescoping. It was imagined that the vehicle would spin up its rotors while still in the aeroshell, then drop to transition to powered flight somewhat analogously to Dragonfly (see Fig. 9.4).

It seems, then, that rotorcraft that are much more ambitious than Ingenuity have fundamental challenges. Whether the value of scientific exploration, or the support for human exploration, merits the investment to meet these challenges remains to be seen. Similar challenges confront fixed-wing or lighter-than-air exploration systems on Mars, and the choice of platform will depend on the mission.

The Ingenuity concept does seem to be proving useful at its current scale, to serve as a scout for a rover vehicle. At the typical pace of rover movement, of the order of a few tens to ~100 m per Sol under good conditions, a scout helicopter that can perform a single look-ahead flight of a few hundred meters has value if flights can be executed every few days; however, should the hazard

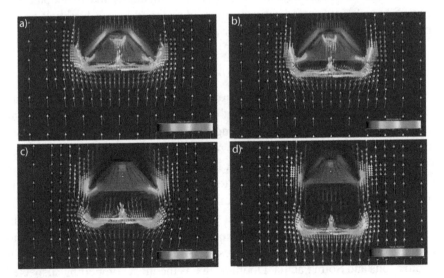

Fig. 9.4 Computational fluid dynamics simulation of a 2-m coaxial rotor vehicle dropping away from a Mars Pathfinder–like aeroshell in vertical descent. As the rotorcraft drops away it passes through a vortex ring state.
Source: NASA.

Fig. 9.5 The Mars surface cruise drone shown by the Chinese Academy of Sciences appears to have many design similarities with Ingenuity. Ironically, the rotor blades appear to lack "Chinese weights."
Source: NSSC/CAS.

avoidance logic on a rover improve to the point where it can drive hundreds of meters, it may start to outpace its aerial scout (which, after all, relies on the rover to perform data relay). It seems likely, however, that we may see Ingenuity-like vehicles again, if only for public relations, and especially when human exploration begins. Interestingly, there have been indications that future Chinese Mars missions might also incorporate an Ingenuity-like helicopter (see Fig. 9.5)[8]—a "Mars surface cruise drone" developed by the National Space Science Center of the Chinese Academy of Sciences (NSSC/CAS).

TITAN

The discussion in the previous section is a natural consequence of the planetary environment, namely that flying on Mars is hard. For the same reason that Dragonfly emerged as a compelling means to address scientific questions on Titan, the possibilities for future rotorcraft there are legion. Compared with a conventional rover, a rotorcraft like Dragonfly offers an order of magnitude more range performance, able to fly a couple of hundred kilometers in total; however, this remains only regional, not global, mobility.

Just as NASA has sent Mars rovers to different locations, it is straightforward to imagine that Dragonfly-like vehicles could be sent to other locations on Titan. One obvious target,[9] of interest geologically and astrobiologically, is Doom Mons and Sotra Facula. Doom Mons‡ is the most compelling suspected cryovolcano on Titan, with a steep pit adjacent to a mountain and some appar-

---

‡Mountains on Titan are named after mountains in the works of J. R. R. Tolkien.

ently extruded material (perhaps liquid water, maybe with some ammonia antifreeze) having flowed away from it, forming the bright region Sotra Facula. There is no guarantee that the extruded material is water—it might be some sort of remobilized organic material, somewhat analogous to the salt glaciers found in Iran—nor that it came from the deep interior, and arguments have been advanced as to why impact crater melt, to be explored by Dragonfly at Selk, are more promising from a prebiotic chemistry standpoint in terms of how warm the material may have been. However, Doom Mons would be an interesting place to explore. Another interesting target might be the largest crater on Titan, Menrva, some 400 km (250 miles) in diameter (see Fig. 9.6). Such a large crater may have excavated deep into the crust, perhaps breaching it to expose ocean material underneath.[10]

Both Doom and Menrva are at moderate latitudes (14 deg S and 20 deg N, respectively), but are on the opposite side of Titan from Selk. This not only places them far out of reach of Dragonfly, being some 5000 km away, but also on the leading hemisphere of tidally locked Titan. This means that a spacecraft arriving from the inner solar system aimed at these targets must enter Titan's atmosphere at a rather higher speed than Dragonfly, which hits the receding face of Titan; therefore, Titan's orbital velocity around Saturn is subtracted from the arrival velocity. For Doom or Menrva it would be added, resulting in an entry speed of perhaps 15 km/s (33500 mph) instead of 6 km/s. This would require a redesigned (and heavier) aeroshell, likely with a higher-performance thermal protection material able to tolerate the higher loads. But the rotorcraft vehicle itself could be essentially unmodified.

The most frequently suggested target for further exploration on Titan (in this author's experience) is its seas. The most interesting aspects of these are

Fig. 9.6 The Menrva impact structure (center) would be interesting to explore. A few dark streaks are seen nestled near the rim of the crater and to the west. River channels are apparent to the southwest of the crater and to the east.
Source: Author.

probably the straits that link the major basins. First is Trevize Fretum,[§] which links Ligeia Mare to Kraken Mare (see Fig. 9.7).[¶] The second is Seldon Fretum, which connects the two major basins of Kraken Mare. Some theories suggest the basins may have slightly different compositions, with the northernmost (Punga and Ligeia) being "fresher" or more methane-rich, and the equatorward Kraken being "saltier" or richer in ethane, the result of a balance of rainfall and evaporation, with the latitudinal gradient of these effects being moderated by mixing from tidal currents, which may be relatively strong in the straits.[11] Thus, as well as offering doubtless strikingly familiar coastal landscapes, this region would be of great interest from an oceanographic perspective.

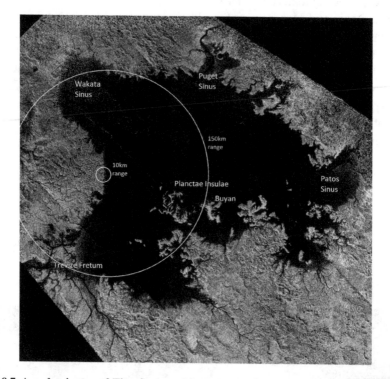

Fig. 9.7 A radar image of Titan's second-largest sea of methane, Ligeia Mare, with a number of its named features labeled. Dragonfly as presently designed could in theory fly out about 10 km from shore (circle and return). A vehicle with a range of 150 km could traverse all but the central region of the sea, although the terrain around the south and east especially is very rugged, so a healthy range margin should be incorporated to find a safe landing spot.
Source: Author.

[§]Freta, or straits, are named after characters in the Asimov *Foundation* series.
[¶]The three seas, Kraken, Ligeia, and Punga, listed in order of decreasing size from 1000 km to 300 km, are named after mythical sea monsters.

Although the seas could be explored by one or more drifting buoys [essentially what the proposed Titan Mare Explorer (TiME) capsule was], a propelled surface vessel (a boat, perhaps with expendable dropsondes, like instrumented depth charges), or even a submarine or a hovercraft, a rotorcraft is also a feasible platform.

A minimum modification to Dragonfly could be simply to adapt its instrumentation and sampling system. The existing mass spectrometer would be well able to analyze the sea composition, if the liquid could be ingested somehow. It is easy to imagine a sampling hose dangling from a hovering vehicle—such arrangements are used in terrestrial helicopters for forest fire–fighting operations. Similarly, a dipping sonar transducer could be lowered into the liquid to measure the depth of the sea, just as they are dunked by antisubmarine helicopters on Earth. It is easy to imagine a simple substitution of the Dragonfly seismometer winch with a longer cable for a sonar, and sonar transducers capable of cryogenic operation were tested during the TiME development study.[12] With such modest adaptations, a Dragonfly vehicle could make sorties a few kilometers from shore, returning landward to perform recharge and data downlink.

With unmodified mobility capability, however, such aerial explorations would be limited to the margins of the seas. An obvious, but significant, adaptation of the Dragonfly vehicle would be to add floats, much as on many terrestrial helicopters.** Some consideration would need to be made for such issues as the generation of seaspray by the rotor downwash and of course the stability of the vehicle against wave motion or even running aground.[13] Beyond stability against capsize, if direct-to-Earth communication is to be performed, the antenna gimbal must be able to slew fast enough to compensate for the vehicle motion in the waves (or an electronically steerable phased-array antenna used instead). Assuming these issues can be addressed, an amphibious Dragonfly could traverse the seas with ~20-km (12 mile) hops, mapping compositional variations and seafloor depths. Note that if such a mission were to be contemplated without an orbiter capable of data relay in support, it would need to wait until the 2040s, when it is Titan northern summer and thus the Earth is visible from the seas.[14]

A Dragonfly-like vehicle could be scaled up (although not within the strictures of the present New Frontiers program), perhaps with a more capable radioisotope power source such as one with a more efficient Stirling generator rather than a thermoelectric converter. A larger battery would allow longer flight range or greater altitude capability for profiling flights. The design space is rather unconstrained-many viable designs can be imagined.

As had been recognized before Dragonfly, small battery-powered rotorcraft could (if the thermal control system is designed appropriately, as in the

---

**In fact the first amphibious helicopter with floats was the Vought-Sikorsky VS-300 which made a flight from the water in 1941. This single-seat single-main-rotor helicopter, an immediate predecessor of the R-4 shown in Fig. 1.5, initially used two lift fans on the tail instead of a cyclic control.

Bumblebee study) perform a "daughtercraft" reconnaissance or sample retrieval role for a larger, less mobile platform such as a balloon or fixed lander. The reconnaissance role is essentially that being performed by Ingenuity today, but the extension to sample retrieval would be a useful capability. This would introduce some interesting challenges in the sample acquisition system, in that the weight-on-bit would be necessarily small, even if the vehicle were commanded to apply downward thrust with its rotors while drilling.

NASA has a canonical progression of mission types to planetary targets of increasing complexity and capability: flyby, orbit, land, rove, sample return. The NASA Innovative Advanced Concepts (NIAC) program recently funded a study of a Titan sample return mission.[15]

Such a mission would have formidable propulsion requirements, to bring a capsule containing Titan samples all the way back to Earth from the Saturnian system. The trick to making it manageable would be to use Titan material as rocket propellant.[††] Liquid methane can simply be condensed from the air, just as water moisture can be harvested on Earth by appropriate compression/ expansion or refrigeration of the air. There is no need to land in the methane seas. The oxidizer to burn the methane fuel in is rather harder to make—water-rich ice material must be excavated, melted, and electrolyzed to yield oxygen, a rather energy-intensive process. Calculations[16] show it might take a couple of years with a modest radioisotope power source to manufacture the 2–3 t of liquid oxygen and methane needed for the return trip. Conveniently, the ambient Titan temperatures are suitable for storing these propellants in liquid form without requiring refrigeration or heavy pressure tanks.

Interestingly, although Titan's gravity is rather low at 1.35 m/s², the dense atmosphere imposes a severe drag penalty on the rocket ascent trajectory. Launch into space is most efficiently done quickly, because every second in which the vehicle is climbing is another 1.35 m/s of gravity penalty[‡‡]. However, accelerating quickly to reduce the gravity loss means incurring a large aerodynamic drag loss instead (and perhaps an indirect performance penalty in having to strengthen the vehicle hull to tolerate the dynamic pressure). Thus, in fact, it may be that the first part of the ascent is best performed by means other than simple rocket propulsion, using aerodynamic or buoyant lift to climb above the bulk of the atmosphere. The drag and gravity losses could be reduced significantly by starting the rocket ascent at an altitude of 30 km or more. At 50 km, the ambient density is only 0.3 kg/m³, compared with 5.4 kg/m³ at the surface. The 0.3 kg/m³ is a density at which it is known that rotors can work: The world-record helicopter flight altitude on Earth [a Eurocopter AS350 Écureuil at 42,500 ft (12,954 m) in 2002] was performed at this density altitude. In fact,

---

[††]The value of Titan as a refueling stop in the outer solar system was recognized by Arthur C. Clarke in his 1975 novel *Imperial Earth*.

[‡‡]It is usual to express rocket propulsive requirements in mission design as a velocity increment or "Delta-V".

the altitude limitation may be more a function of engine performance than rotor aerodynamics. In any case, 50 km seems an achievable altitude on Titan.

A rotorcraft first stage might be more straightforward than a balloon, although that is possible, too. If rotors are installed on the vehicle, they could (as for Dragonfly) be used for initial landing and perhaps relocation on the surface. Thus, there are many intriguing possibilities for future rotorcraft on Titan.

BACK TO THE FUTURE

It is interesting to reflect that both urban air mobility (UAM) and planetary aviation are in some ways where aeronautics as a whole was in the first decades of the 20th century. The UAM domain is seeing a Cambrian explosion of ideas—nobody quite knows what the right formula is for providing safe, quiet, and efficient vertical takeoff and landing (VTOL) transport with electrical or hybrid power. There are designs and demonstrators of everything from multirotor vehicles to tilt rotors, hybrid vehicles with lifting rotors and wings, and airplane-like vehicles drawing short takeoff and landing (STOL) performance with blown flaps.

Just as there was a plethora of weird and wonderful lifeforms at the end of the pre-Cambrian era 500 million years ago, enabled perhaps by an increase in atmospheric oxygen levels, so the UAM designs are exploding in diversity. When the venture capital eventually runs out, the evolutionary pressures of commercial reality will squeeze out the less fit designs, and whatever works best will remain. This is just what happened with the design of tanks and aircraft before and after World War II, with (for example) the low-hulled main battle tank with a single turret and main gun becoming the near-universal archetype.

For planetary aviation, the situation is a little different, in that every mission is ad-hoc. Although there can be some inheritance from one spacecraft to the next (eg, after many weird-shaped satellites in the 1960s, almost all are now box-shaped with rectangular hinged solar panels), planetary missions are typically one-offs, designed starting from a blank sheet of paper. (This was certainly the case with Dragonfly!) Thus, the process of small variation and natural selection does not occur. Further, as this chapter has described, the peculiar situations of each planetary environment demand different design solutions. It is hoped that this book has given the reader the perspective with which to consider them.

REFERENCES

1   For another outlook, see Young, L. A., Lee, P., Aiken, E., Briggs, G., Pisanich, G. M., Withrow-Maser, S., and Cummings, H., "The Future of Rotorcraft and Other Aerial Vehicles for Mars Exploration," *Vertical Flight Society's 77th Annual Forum & Technology Display* 2021, https://ntrs.nasa.gov/api/citations/20210014168/downloads/1473_ Young_042321.pdf

2   For example, Dyson, R., and Bruder, G., "Progress Towards the Development of a Long-Lived Venus Lander Duplex System," *46th AIAA/ASME/SAE/ASEE Joint Propulsion Conference & Exhibit*, AIAA-2011-6917, May 2011. The idea is, like most, rather older—in fact, I recall a presentation on an earlier incarnation at the same meeting where Savu presented a Mars helicopter: Schock, A., "Integration of Radioisotope Heat Source with Stirling Engine and Cooler for Venus Internal-Structure Mission," No. IAF-93-R. 1.426, International Astronautical Federation, Graz, Austria, 1993.

3   Lorenz, R. D., Crisp, D., and Huber, L., "Venus Atmospheric Structure and Dynamics from the VEGA Lander and Balloons: New Results and PDS Archive," *Icarus*, Vol. 305, 2018, pp. 277–283.

4   Landis, G. A., LaMarre, C., and Colozza, A., "Venus Atmospheric Exploration by Solar Aircraft," *Acta Astronautica*, Vol. 56, No. 8, 2005, pp. 750–755.

5   Johnson, W., Withrow-Maser, S., Young, L., Malpica, C., Koning, W., Fehler, M., Tuano, A., Chan, A., Datta, A., and Chi, C., "Mars Science Helicopter Conceptual Design," NASA/TM 2020–220485, NASA, 2020. This report contains a lot of study results of a variety of vehicle sizes and designs making it somewhat difficult to trace a definitive reference design. It seems prudent to consider the results very preliminary.

6   Delaune, J., Izraelevitz, J., Young, L. A., Rapin, W., Sklyanskiy, E., Johnson, W., Schutte, A., Fraeman, A., Scott, V., Leake, C., and Ballesteros, E., "Motivations and Preliminary Design for Mid-Air Deployment of a Science Rotorcraft on Mars," *ASCEND 2020*, p. 4030.

7   Young, L. A., Delaune, J., Johnson, W., Withrow, S., Cummings, H., Sklyanskiy, E., Izraelevitz, J., Schutte, A., Fraeman, A., and Bhagwat, R., "Design Considerations for a Mars Highland Helicopter," AIAA-2020-4027, *ASCEND 2020*.

8   Jones, A., "China Is Developing Its Own Mars Helicopter," *SpaceNews*, 1 Sept. 2021, https://spacenews.com/china-is-developing-its-own-mars-helicopter/

9   Lorenz, R. D., *Saturn's Moon Titan: Owners' Workshop Manual*, Haynes, Aug. 2020.

10 Crósta, A. P., Silber, E. A., Lopes, R. M. C., Johnson, B. C., Bjonnes, E., Malaska, M. J., Vance, S. D., Sotin, C., Solomonidou, A., and Soderblom, J. M., "Modeling the Formation of Menrva Impact Crater on Titan: Implications for Habitability," *Icarus*, Vol. 370, 2021, https://doi.org/10.1016/j.icarus.2021.114679

11 Vincent, D., Karatekin, O., Lorenz, R. D., Dehant, V., and Deleersnijder, E., "A Numerical Study of Tides in Titan's Northern Seas, Kraken and Ligeia Maria," *Icarus*, Vol. 310, 2018, pp. 105–126.

12 Arvelo, K., and Lorenz, R. D., "Plumbing the Depths of Ligeia, Considerations for Acoustic Depth Sounding in Titan's Hydrocarbon Seas," *Journal of the Acoustical Society of America*, Vol. 134, 2013, pp. 4335–4351.

13 Lorenz, R., and Mann, J., "Seakeeping on Ligeia Mare: Dynamic Response of a Floating Capsule to Waves on the Hydrocarbon Seas of Saturn's Moon Titan," *Johns Hopkins/APL Technical Digest*, Vol. 33, No. 2, 2015, pp. 82–94.

14 Lorenz, R. D., and Newman, C. E., "Twilight on Ligeia: Implications of Communications Geometry and Seasonal Winds for Exploring Titan's Seas 2020–2040," *Advances in Space Research*, Vol. 56, 2015, pp. 190–204.

15 Pultarova, T., "Scientists Ponder How to Get Samples from Saturn's Weird Moon Titan," *Space.com*, 18 May 2021, https://www.space.com/saturn-moon-titan-sample-return-mission

16 Landis, G. A., Oleson, S. R., Turnbull, E. R., Lorenz, R. D., Smith, D. A., Packard, T., Gyekenyesi, J. Z., Colozza, A. J. and Fittje, J. E., 2022. "Mission Incredible: A Titan Sample Return Using In-Situ Propellants," In *AIAA SCITECH 2022 Forum*, (p. 1570).

Appendix

PLANETARY AVIATION BY THE NUMBERS

Some empirical ("allometric") relationships for the flight power of airplanes, airships and helicopters in planetary atmospheres are given in Ref. 1. But fundamentally, the most useful relationship in contemplating planetary rotorcraft is that provided by the actuator disk model to indicate the idealized power to hover,

$$P = T^{1.5}/\sqrt{(2 \cdot \rho \cdot A)}$$

P is the power applied to the airstream to generate thrust T, given the fluid density ρ and the disk area A (which equals $N\pi D^2/4$, where D is the rotor diameter and N is the number of rotors). In hover, $T = mg$, where m is the vehicle mass and g the local acceleration due to gravity.

All else being equal, this implies that a vehicle will need a power proportional to the 1.5th power of gravity, and the reciprocal of the square root of density. Substituting values from Table 2.1, it is easy to see why Mars is a difficult place to fly: the cruel effect of the low density greatly exceeds the modest benefit of the lower gravity. Titan wins on both counts.

The practical realities of actual vehicles (where there is profile drag on the rotors, parasitic loads such as power to a tail rotor and the like, and a need for thrust margin to climb and maneuver) is such that the power actually needed[1] may typically be 50–100% higher than this idealization would suggest.

In fast forward flight, the drag on the vehicle will become the dominant power sink. To a first order, the forward flight speed at which this becomes the case can be estimated as that at which the drag equals the vehicle weight. Thus, the rotors must then provide a thrust equal to 141% of the weight.

(Because the weight and drag are equal, the thrust is inclined 45 deg forward.) Per the actuator disk equation, the required power will be equal to about $1.41^{1.5} = 1.67$ the power required for hover. Since weight equals drag in this heuristic estimate, we have

$$mg = 0.5\ S\ C_d\, \rho\ V^2$$

where S is a reference area and C_d the associated drag coefficient. For Dragonfly, $S = {\sim}2.0$ m^2, and given the many appendages and rather blunt nose, we can guess $C_d = {\sim}0.5$. (A very clean streamlined shape might have $C_d = {\sim}0.1$.) Substituting $m = {\sim}800$ kg and 5.4 kg/m^3, we find $V = {\sim}\ 20$ m/s. Clearly, this should be treated only as a crude estimate, but at least it provides a guide.

Perhaps even more fundamental than the power required for a given flight condition is the question of whether a given vehicle can fly at all. You may recall from Chapter 2 that the rotor disk(s) must hurl down a momentum flux of air equal to the vehicle weight. Thus, for rotor area A, the momentum flux is $A\rho v^2$, where v is the velocity at the disk. We may assume that at best this velocity is (averaged over the disk) about one-sixth of the tip velocity, which in turn is limited to about two thirds of the speed of sound c. It then follows that the maximum thrust attainable is $A\rho(c/9)^2$.

Let us imagine a 2-t rocket stage at 40 km altitude on Titan. The thrust needed is 2600 N. Given a sound speed of 180 m/s and density of 0.3 kg/m^3, the required disk area is $(2600/(0.3 \times (180/9)^2)) = 22$ m^2. If we imagine a quad-rotor vehicle, then each disk must have an area of 5.5 m^2 and thus rotors of a radius of 1.3 m^2 (about double that of Dragonfly).

Note that a key metric is the "about one-sixth" above. In wing terms, generating a downwards velocity component equal to a sixth of the incident velocity corresponds to deviating the flow relative to the airfoil by about 10 deg. This also corresponds to a lift coefficient (a characteristic of a given airfoil at a certain angle of attack at some Mach and Reynolds number, and one that is normalized to the area of the wing) of about 0.15, which is not exceptional. A good wing might develop a lift coefficient of better than 0.6 before stalling. However, this two-dimensional framework is a gross simplification for a propeller or rotor, because the relative wind varies along the span (and is generally rather less than the tip speed), and the airfoil may be twisted so that the effective angle of attack varies along the blade. Thus, the rotorcraft industry tends to emphasize a thrust coefficient that relates the overall momentum flux (ie, thrust) to the other parameters such as tip velocity.

REFERENCE

1   Lorenz, R. D., "Scaling Laws for Flight Power of Airships, Airplanes and Helicopters: Application to Planetary Exploration," *Journal of Aircraft*, Vol. 38, 2001, pp. 208–214.

INDEX

Note: Page numbers followed by f indicate figures and those followed by a t indicate tables.

SUPPORTING MATERIALS

A complete listing of titles in the Library of Flight series is available from AIAA's electronic library, Aerospace Research Central (ARC), at arc.aiaa.org. Visit ARC frequently to stay abreast of product changes, corrections, special offers, and new publications.

AIAA is committed to devoting resources to the education of both practicing and future aerospace professionals. In 1996, the AIAA Foundation was founded. Its programs enhance scientific literacy and advance the arts and sciences of aerospace. For more information, please visit www.aiaafoundation.org.

CPSIA information can be obtained
at www.ICGtesting.com
Printed in the USA
LVHW050745270123
737542LV00007B/1

9 781624 106361